Blood Alcohol, Breath Alcohol, Impairment and the Law

A manual for law enforcement, attorneys, and others

Interested in alcohol issues in law enforcement today

Alfred W. O'Daire, Jr.

Diplomate, American College of Forensic Examiners

Board Certified Forensic Examiner

Member, American Academy of Physician Assistants

Board certified Physician Assistant

AuthorHouse™
1663 Liberty Drive, Suite 200
Bloomington, IN 47403
www.authorhouse.com
Phone: 1-800-839-8640

First published by AuthorHouse 3/13/2009
First edition September 2008, Colonial Heights, Virginia

ISBN: 978-1-4389-2580-6 (sc)

Library of Congress Control Number: 2008911434

Printed in the United States of America
Bloomington, Indiana
This book is printed on acid-free paper.

This text is dedicated to

William F. Binford, Jr., Esq.

A true southern gentleman
Billy always made you think, and was directly responsible
for guiding me to put this information to paper.
He was a great friend, a great Attorney, and a great man taken from
us as far too young of a man. He may not be here physically, but
he is certainly with us all in spirit. The lessons he taught us all
will serve each of us well for the rest of our lives.

Table of Contents

Charts

Forward

It is important that all of us involved in the hunt for, arrest of, prosecution of, and punishment of DUI suspects remember that not everyone arrested for any crime is guilty, not everyone seeking a vigorous defense and to be exonerated is innocent, and a fair trial with the best possible defense and fairest prosecution is the best situation.

If we all keep the above statement in our minds and hearts, then justice will always be served. It is not unreasonable for law enforcement officers to make an arrest, and find during trial that there really is a potential and honest defense that will prove the suspect is innocent. No officer or prosecutor should ever find it offensive that a subject is released by the court.

Every key person in the detection, arrest, and trial of a DUI suspect has a specific and important role. So long as each of these persons perform their duties as honestly and sincerely as possible, then no innocent suspect will be convicted, and no guilty suspect should go free.

It is better to loose an occasional suspect to an acquittal when they may truly be guilty than to convict someone that is not guilty of the crime with which they had been charged.

This text is produced in honor of all of the Law Enforcement Officers, all the Prosecutors, and all the Judges that I have come into contact with over the past fifteen years of appearing in court. There are several defense Attorneys that I have become quite close to over the years, and to those gentlemen and ladies, I thank you for allowing me to serve the public in what was sometimes a very difficult position.

There was always an amazing amount of professionalism and integrity on behalf of all these members of the judicial system. I hope you all continue to arrest, prosecute, and punish those that are actually guilty of DUI. You are making a difference and are making the streets of Virginia, and the United States, a much safer place. You must keep up the good work you are all doing, and continue to fairly dispense justice to the best of your ability.

Hopefully, there will be others to come along and ply their trade in science to help the law enforcement community and judicial system to understand more and more the ins and outs of the science behind DUI enforcement. There is considerable harm to be done by convicting an innocent suspect as well as to release the guilty without the proper punishment. Rehabilitation is a wonderful concept, but in order to make permanent changes to a person's behavior, it must be painful in order to ensure it will last. It is unfortunate that we as human beings, must be trained like any other animal—if it hurts, we are less likely to repeat a behavior.

Alfred W. O'Daire, Jr.

Preface

First and foremost, I think it is important that I make it clear who I am and why I am qualified to provide this information about such an important issue as blood alcohol and driving under the influence. I have had the wonderful fortune of several careers in my lifetime, and two in particular have prepared me for the task undertaken in writing this text.

From the time I was a very young child, I have wanted to be a police officer. Even when faced with the problem of being an 18 year old high school graduate, and not being old enough to be a police officer, I joined the military with the intention of passing time until I was old enough to become a cop.

My first dealings with enforcing the law came while on active duty in the Marine Corps. I was stationed in San Diego, California, and was assigned to Headquarters Company in a ceremonial unit. As luck would have it, the Military Police unit was also part of the Headquarters Company, and a sudden need for a Staff Sergeant arose one weekend during the last year of my enlistment. I had nine months to go and my entire command knew I was getting out and was going to be a cop somewhere. I was the logical choice (in the eyes of Marine Corps, at least) to be laterally transferred to the MP unit, and overnight, with no training, no experience, and no further notice, I became a Military Police Supervisor. What a change for a good old fashioned grunt—I was an infantryman by training even though I had been in a ceremonial unit since my return from Southeast Asia.

I was told of the move when I arrived at my regular work station on Monday morning, and went to the Provost Marshall's office only to be told I was late, and my shift was "on the street". I was given an armband, a well worn .45 pistol and web gear, a portable radio, and the keys to the supervisor's pickup. The rest, as is sometimes said, is history. I was hooked on the duties and responsibilities of enforcing the law, and even if it was on a very small base with a maximum speed limit of 10 MPH, I was finally doing what I had dreamed of since watching Broderick Crawford on the old show "Highway Patrol", and Reid and Malloy on "Adam 12".

I left the Marine Corps, and went to work for the San Diego Police Department. I worked with a squad known as the "enforcement" squad. This group of officers had the sole responsibility to enforce the drunk driving laws all around the city. It was a very specialized unit that was on the road from 8 PM to 4 AM, seven days a week, and did nothing but look for drunk drivers. On any given night, we were out there, watching every car around us looking for any telltale sign of impairment. The training was certainly "on the job", and I was fortunate enough to have a grizzly old timer named Jim Smith for a partner and teacher.

At first, Jim wasn't too thrilled about having a rookie partner, and was actually getting ready to retire after 30 years with the department, but he put up with me knowing it was only for short time. Little did I know then that Jim was going to teach me the basis for the work for the rest of my life and the bizarre direction these lessons would lead. I learned to spot a drunk driver from the most subtle sign. I learned how to perform some roadside tests in order to solidify my case for court if I decided to make an arrest or to release the driver and not waste valuable hunting time on a driver not impaired enough to be dangerous.

In San Diego, anyone arrested for DUI at that time was provided the option of a breath test, a urine test, or a blood test for determination of their blood alcohol level after arrest. We were fortunate in having a full time lab right in the police headquarters, and breath or urine provided a result within moments. Blood samples took a little longer, so we naturally attempted to sway a suspected drunk driver into the breath or urine test. This was, of course, for our convenience, and had nothing to do with justice. It didn't matter to us, because we were told they were all accurate and equal tests for determining blood alcohol levels. We were expected to produce two DUI arrests apiece during each shift, and the less time spent at the station processing the suspect, the better.

After a little more than a year in San Diego, the urge to move home to the east coast got the best of me, and I relocated back to Virginia, where I had grown up. The Petersburg, Virginia Bureau of Police was the first department to call me for an interview and make me a job offer, and I couldn't wait to start work. The sergeant in charge of the small traffic squad knew me from my days in the area before I joined the service, and he and I spoke about the possibility of being assigned to the traffic division after the academy. It seems that in Petersburg, an assignment to traffic was somewhat of a promotion for a patrol officer, but he assured me he would watch my work, and if I was as good as I told him I was, I would get a fair shot at a slot in traffic should one come open.

In the days of the mid 70's, the VASAP (Virginia Alcohol Safety Action Program) office was providing overtime pay for three officers every Friday and Saturday night to do DUI patrols. Two officers worked the street, and one stayed in the office and kept the Smith and Wesson 900A Breathalyzer warmed up. For four hours each of these nights, two off duty officers worked for overtime pay just hunting for drunk drivers. This was right up my alley, and with an annual salary of only $10,000, every penny mattered. I quickly became a favorite of the traffic supervisor for these extra shifts, as instead of the regular one, and occasionally two DUI arrests, I would make three or four every four hour shift. The training in San Diego was beginning to pay off.

After about six months, an opening came available on the traffic squad, and even though the sergeants had recently changed, I was still given the slot above several much more senior officers. This allowed me to attend several specialized schools about traffic accident investigation, hit and run investigation, and drunk driving. I was fortunate to get a slot in the state forensic lab's next breathalyzer operator's course. I became one of the departments leading DUI arresting officers of any time.

I was able to help develop the DUI detection abilities of the other officers by teaching at the recruit academy as well as the required in service academy all officers had to attend every two years. This further honed my skills for DUI detection and allowed me to rack up a tremendous number of DUI arrests with a significant conviction rate. I lost very few DUI cases in my short police career.

Unfortunately, after five years of being a police officer, I was injured on the job (making a DUI arrest during a VASAP shift). This injury led to my medical retirement from law enforcement. Since this was the only profession in which I had ever had any interest, I was devastated and had no idea where my life could possibly lead. I had just become a father for the first time and had a young son counting on me to provide for him. I can remember sitting in the surgeon's office, and telling him he had to clear me back to work. The advice he gave me that day would change my life forever. It was the same advice I was given years before by my father and other adults with a significant role in my development, but this time I actually listened. Timing is everything. He told me to go to college and learn a new job. He told me I'd never be able to be a cop again.

All the air was out of my balloon this time. I was so crushed by the news that I could no longer be a cop that I really and truly had no idea where I was going to go next so I enrolled in college. First it was the local university just because it was close to home, cheap, and I was bored to tears. After three semesters of finding out how much I really enjoyed learning, I began to look for a bigger school with a great science curriculum, because I had fallen in love with science. I began to see correlations between the work I had done on the street investigating accidents and making DUI arrests, and the science in the textbooks. I was starting to understand the part of my previous work that was so mysterious before. Biology, chemistry, and physics had real meaning from my real world experiences of my past life. It was much more than just textbooks and theory. It had real world applications and meanings, and made it appear as if a light bulb suddenly came on. Each successive class was more intense and made more sense of my meager science background from before college.

After college, armed with a bachelor's degree, I found there was very little that could be done at that education level, so I began to apply to medical schools. During this arduous process, I found out about the PA profession. It seemed that being a physician assistant would allow me to make more money than a lab scientist with a bachelor's degree, and would allow me to practice medicine, without the time investment of four more years of medical school, and then three or four more years of a residency. I found my place in post-law enforcement life, and got accepted to the PA program in Baltimore, Maryland.

After two more years of school, I began to practice family medicine and emergency medicine. I was back to helping people that were in need, and was actually making enough money at it now to support a young, growing family.

A gentleman (Mr. William F. Binford, Jr., Esq.—"Billy") with whom I had become very close friends was a very successful attorney in central Virginia. At this time Billy had actually narrowed his legal practice to DUI work only. We originally met when he was defending clients that I had arrested while working the streets of Petersburg. He called me up one day and asked me to come by his office to talk about a case that seemed to be causing him some concern about the science behind the breath test. I agreed to meet with him, and the question that day was the first step on the journey that has led me to this point and the production of this text. Billy was interested in the science behind the DUI laws and the tests and how it was applied out of concern that there may be some room for error in the process. I questioned him many times as a police officer about defending people that I considered guilty (or I certainly wouldn't have arrested them) and I continued to question him as a scientist, although I now understood more about the possibility of errors on the part of the state and the assumptions necessary to use the tests involved. His answer was always the same—everyone deserved the best defense the law allowed, and it would be wrong to not provide that defense to the best of our ability.

Many times over the next few years, Billy and I worked together on many cases, some involving very simple scientific points and occasionally some with very complex scientific issues. Word got around, and other attorneys called for consults and to retain my services for similar matters. I was able, through my medical practice, to attend several specialized classes involving drinking and drug impairment. I developed and coordinated several research protocols involving drinking and variations in blood and breath alcohol levels. I began to see variations in the use of the "standardized field sobriety tests" and realized that many law enforcement officers, prosecutors, defense attorneys, and even judges didn't understand these tests and how to use them to draw reasonable conclusions about impairment. They all certainly understood that the manuals said they were effective, but the necessary details were very frequently lost in the interpretation, and false conclusions abounded.

I began to lecture in different venues about the science behind drinking, and impairment. I provided text materials and lecture time to the state bar association continuing legal education conferences on DUI defense here in the Commonwealth of Virginia. I began receiving telephone inquiries from all over the United States in reference to my services on providing consultations and testimony on the science behind alcohol and impairment. I was able to provide service to courts in all types of cases, including homicides, rapes, and various other criminal and civil matters involving issues of impairment. I was even consulted on and provided expert testimony on cases involving work place matters dealing with alcohol and possible impairment issues. The civil matters involved the normal accident and injury cases as well as an occasional divorce or child custody case. I have even been consulted on one case in which a Catholic annulment hinged on the groom's drinking history of over twenty years. This case went to the highest judge on Earth to be decided in favor of the client that retained my services.

There are volumes of research documented about DUI, impairment, methods to test for impairment, and even about the standardized field sobriety tests. There is no one place where all of this materiel can be brought together in a sensible text to allow a review and provide an education about the science involving drunk driving. I have attempted to provide that text here. There may be some portions of the following text that will not be interesting or even relevant to one particular case or issue, yet I believe all of the information here is necessary to provide a full understanding of DUI enforcement. This is such a scientifically based offense, that it is necessary to start at the beginning. This involves some high school and college chemistry that is by no means difficult to understand, but is paramount to understanding why certain issues are so very important to the enforcement of the drinking and driving laws of our land. There is simply no way around the basics whether your particular interest is in the evidential breath tests, the field sobriety testing, the absorption and elimination of alcohol, or using this information for the extrapolation of blood alcohol curves.

Hopefully, this will all tie together in a manner that will make good sense to provide the understanding required when a DUI case is up for review or trial. For a defense attorney to understand the charge and the best way to provide for a proper defense, some science is important. If not, there may be no way to determine if a client is telling the truth about his experience, or just trying to undo the inevitable. DUI defense attorneys are becoming so scientifically adept at providing a defense, the police officers and prosecutors of today need to be able to understand all of the issues in order to better overcome any smokescreen type of defense, or to see when it may be in the best interest of justice to not pursue a charge, even when properly placed. It is no longer enough to prosecute a case knowing only the Evidential Breath Test is high enough when the defense may provide significant scientific rebuttal. Consequences of drunk driving convictions are so severe in today's society that it is absolutely imperative that anyone convicted of DUI be guilty of having committed that crime. On the same vein of thought, consequences of driving impaired are so significant that it is also absolutely necessary to remove every impaired driver's ability to cause injury, death, and property destruction by ensuring the appropriate convictions.

The purpose of providing this text is not to allow any drunk driver to go free. With the publicity being made in today's society about alcohol safety, there is no reason for anyone to ever drive while impaired, and anyone that does this should be punished to the fullest extent available. I want nothing to do with allowing anyone to beat a valid DUI charge, and in fact, have refused to testify on behalf of many cases in which I was consulted when it appeared that the science was certainly in support of the state's case. The most significant of these cases was one client that changed his story the day of his trial. Anyone that drinks to the point of impairment and then operates a motor vehicle deserves everything the court can give them. With all the media coverage about drunk driving and the potential for injury and death, only someone with either a severe drinking problem or a lack of concern for fellow citizens would get behind the wheel after drinking. There is no excuse, and anyone that commits the crime should not only be punished, but should be removed from the driving public.

Hopefully, by showing my background, it is easy for the reader to see that I am not interested in helping anyone "beat" an appropriate DUI charge. I believe that all of our law enforcement officers are interested in doing the best job possible. I do not think there are any officers that would purposely convict a single innocent person of any crime. I have been questioned over the years by opposing attorneys, by police officers, and even an occasional judge about why I do so much defense work. It's an easy question to answer. The state has their own experts on payroll and has no reason to pay me to come to court and argue for their case. The defense has no one to argue the science on their behalf. I am frequently asked by prosecutors how often I have testified for the prosecution, with the obvious desire to paint me as a hired gun for the defense. Some courts do not allow the question, but others do nothing to stop the obvious personal attack. My answer is always the same—if the state was to consult me (and they have on at least one occasion), I would be happy to evaluate the matter for them and provide my expert opinion. On the single occasion (up to the time of this writing) that the prosecution has asked me to evaluate a matter I declined all payment of fees and expenses and performed the analysis as a public service for no fee. The Commonwealth Attorney that used my services provided my written opinion to the Defense Attorney and received a guilty plea to manslaughter without so much as an agreement on sentencing.

I don't go to court on every case. I am very picky about the defense cases I agree to work with. Even then, I am not an advocate for the defense—I am an advocate for the scientific truth in the issue at hand. My only purpose is to assist the trier of fact to understand the science involved in a particular issue and make the best decision in a manner that would be fair to the defense and the prosecution. My opinion would be the same when based on the same facts, whether I were brought before the court by the prosecution or the defense. The facts in a case are the facts, and they should not change depending on the side that presents them. I make certain that every client that is referred to me by their attorney understands that fact. No one wins in a case where justice is not served, whether it may be an acquittal or a conviction.

Chapter One

Introduction to Alcohol Forensics

Driving under the influence of alcohol is one of the most serious social and criminal problems in our land today. It is a criminal act with consequences that touch lives of thousands of people in our country, and even in our world, today. It is probably the most common criminal act that the normal, law abiding citizen is likely to be touched by in their lifetime. It is so prevalent that there is a phenomenal chance that nearly everyone will have a family member, close personal friend, or the family member of a close personal friend either arrested for DUI or be injured or killed by an impaired driver. Enforcement of the DUI laws is paramount to our society and keeping the normal citizenry safe from injury and death at the hands of drunk drivers. It is a mandatory function of law enforcement in society today, and, unfortunately takes up entirely too much time for officers that can better be assigned other necessary duties. Due to the prevalence of alcohol and automobiles in our country, law enforcement officers must devote a significant amount of duty time and training time to DUI enforcement activities.

As one of the most common criminal offenses encountered in law enforcement today, the evolution and training of DUI detection and enforcement has been an ongoing process for many years. In the early 1970's, major moves to enhance the enforcement of DUI laws in the United States began to take on an aggressive life of their own. The different states, as well as the federal government, began to recognize the danger of driving under the influence, and the need for more intense enforcement and the training needed to affect that intense enforcement. Deaths, injuries, and major property losses were increasing at an alarming rate every year.

It is obvious that the impaired drivers need to be removed from the driving population, and laws have been passed and updated on an annual basis to make enforcement and punishment easier for our law enforcement officers and courts. If a driver cannot be made sober and safe for all future motor vehicle operation, he or she must be removed from the driving population, either by court order, DMV directive, or if necessary, incarceration. This is not only necessary, but would be irresponsible on the part of government to not provide this service to the everyday driving populace in America. There should be no mistaking the information provided in this text as not being produced in order to thwart law enforcement from making the necessary, and in fact mandatory, arrests for this heinous offense. No drunk or impaired driver should go unpunished. With the significant media coverage of this problem, there is absolutely no excuse for anyone to attempt to operate a motor vehicle today while impaired.

As more and more research energy and money was directed at the problem of impaired driving, it became painfully obvious that this legal issue was steeped deeply in science. This was one of the first criminal offenses that received the scientific community's attention in order to allow more effective policing and enforcement. DNA testing for criminal purposes was pre-infancy at this time, as were most other categories of criminal sciences.

This problem of DUI required that law enforcement officers be able to detect the offender on the street in order to initiate contact in a legal and appropriate manner. After the initial contact, officers needed some method of evaluating the level of impairment in order to make a decision to arrest or not to arrest in order to most effectively remove the dangerous driver from public, and protect the others on the highway. To make the incorrect decision at this point could endanger the public, or incorrectly make an arrest of an innocent citizen. After the decision to arrest, there needed to be some type of mechanism in place in order to determine the level of impairment in a manner easy to document and present as hard, firm, and hopefully, indisputable evidence in order to allow the court or jury to make the correct decision and convict the offending suspect when appropriate. Then, and only then, could the process of punishment and reeducation of the drunk driver be started and a safe and sober driver then be returned to the driving public.

All of these steps were surrounded in science. Not simple, grade school science, but complex chemistry, physics, biochemistry, toxicology, pharmacology, and physiology. These are all college level and graduate school level studies, and medical school disciplines that need extensive scientific backgrounds to understand. Few law enforcement officers, attorneys, and judges have the necessary scientific education and background to allow an understanding of the processes involved in the enforcement of DUI laws. Frequently it was necessary for these major players to depend upon the scientists employed by the government or state labs, or the manufacturers of the alcohol testing equipment. It placed the defense, prosecution, and court at the mercy of a predictable one sided view of the science involved in determining impairment. None of these scientific viewpoints is inaccurate, but could possibly be incomplete. It is highly improbable that this is an intentional error, as these scientists are well trained, and well versed in their particular fields. It is much more likely that it is simply the view that is most likely to produce the best outcome in the easiest and shortest, most direct route. There is little need to expound on information if it is not brought into question. After all, science really is not always hard and fast, and there are usually more than a single explanation for certain facts and issues. Hopefully, this will become quite clear as the reader delves further into this text.

The federal government began their involvement in developing the standardized field sobriety tests in the early to mid 70's. Years of scientific research went into developing these tools for the everyday street cop to use in order to affectively enforce the impaired driver laws of our country.

They had been developed over the years to become a very useful tool that is used on a daily basis across the United States by local, state and federal law enforcement officers tasked with enforcing DUI laws. This development took place over a period of several years by a team of highly acclaimed scientists, using a multidiscipline team of experts to develop, evaluate for effectiveness, and present the recommended tests to the government sponsors and the law enforcement community.

Government supported and privately or industrially backed research has progressed to develop some very useful tools to allow quantitative measurements of alcohol levels in blood and breath specimens. This testing equipment has matured into very effective and accurate instrumentation that is useful, even though there may be some scientific limitations on these devices. When the author was a young street cop, the accepted field alcohol detection device was a clear glass tube with chemically active crystals that changed color when exposed to certain chemicals, one of which was alcohol. Today, officers have such advanced equipment that actually gives a reasonably accurate reading of the suspect's actual breath alcohol level, again with certain scientific limitations. Even though we were taught that these methods were accurate and gave us a true indication of alcohol levels, these tests would never stand the test of being scientifically valid today. The methods available to law enforcement officers today are so far advanced beyond ours of the 1970's that it makes our old methods appear prehistoric in scientific validity and accuracy. This is but one small example of how far the science of DUI enforcement has progressed over the last thirty years.

The detecting, apprehending, testing, bringing to trial, and punishing of DUI offenders is a ballet of a group of players, all with a specific place in the team. The goal of everyone on this team is to remove and punish unsafe, impaired drivers. Everyone involved has a particular role, and everyone needs to work together while doing their own individual parts in order to bring about the proper outcome.

In the scheme of bringing these offenders to justice, the first player to participate is the law enforcement officer. It is mandatory in today's legal climate that the officer involved in detecting and apprehending DUI suspects be well versed in detecting these drivers. He or she must have excellent skills and high levels of knowledge about the standardized field sobriety tests, the use of and limitations of field testing instrumentation, and the evidential breath tests used in their jurisdiction. They must be able to articulate the facts and their conclusions fully in front of the court in order to bring about the proper conclusion to the case. Any failure on behalf of the officer may jeopardize the outcome of the case, as it would allow questions as to the validity of the conclusions.

The next two cast members are the attorneys, one for the prosecution and one for the defense. The consequences of a conviction for DUI today are adequately severe that it would be foolish for anyone to make an appearance on this charge without benefit of legal counsel. Years ago, it was not uncommon to see a defendant appear in court on one of these charges, even after previous convictions, and sign a waiver of legal counsel. This is extremely rare now, and most courts would certainly make every effort to discourage this practice. Most judges will not permit a defendant to handle their own DUI defense without discussing the issue with an attorney first.

Defense attorneys are usually faced with the defendant some time after the arrest, and the potential client normally is adamant about their innocence. Sometimes this is due to lack of understanding about the fine points of law and what may or may not constitute an offense. Other times, it may be that the client withholds certain facts from their story, in hopes an attorney may present a viable defense and they may get away with something. Yet others may actually have what may eventually be a good defensive stand with valid scientific and legal merit for going into court and making the appropriate presentation. It takes a well versed defense attorney with a good background on the science behind DUI enforcement in order to determine where a particular client may fall in this scheme. It is quite a complex issue to put together the full case and prepare for the trial and presentation of the scientific expert testimony necessary to fully defend a case. The properly prepared defense attorney understands the issues from a legal standpoint, but has also made considerable effort to understand the scientific issues enough to determine when a forensic evaluation may be necessary. They should also be able to see when the client may be withholding certain facts.

Many times, the prosecuting attorney has little time to prepare for a DUI trial when called for the first time. Unless it is a special case, usually a high profile case, or one in which there are long term consequences depending on the outcome, most prosecutors do not become involved in the run of the mill DUI case until the day of trial. They must fully understand the facts and the basis for the stop, arrest, and test results in order to protect the interest of the state and the public in court. A good prosecutor, knowledgeable in all aspects of the science of alcohol metabolism and testing, will know when a case should be tried with maximum effort and aggressiveness, and which cases would be best served to plead away to lesser charges or even not pursue to any conclusion. It certainly should not be the goal of a prosecutor to obtain a conviction of every case brought by the police at all costs although many prosecutors feel if they do not aggressively seek a conviction in every case they have failed. Sometimes it may be in the best interest of justice to not convict on a charge. This does not mean there was a mistake on behalf of the arresting officer because, fortunately, there is a drastically different standard when looking at probable cause to bring charges and guilty beyond a reasonable doubt.

On some cases, particularly those that hinge on a certain scientific issue, there may be expert witnesses involved. Whether appearing on behalf of the defense or the prosecution, the goal and purpose of scientific experts is to enlighten the trier of fact as to the scientific issues as they pertain to that case. A defense expert is not in court as an advocate for the defense. Yes, they are certainly being paid by the defendant, but they are being paid for their time, not their words. A good scientific expert will testify to the facts in general and how they apply to a particular case. A defense attorney will hire an expert only if the facts are present that will support the argument of innocence for his client. A defense expert is not an inexpensive venture. It is unlikely that a good defense attorney will recommend to a client that he or she hire an expert unless they truly believe there is a valid scientific argument. In addition, most good defense experts will not risk their reputation appearing on a case with a questionable argument. Most defense experts appear on a very small percentage of cases in which they are consulted, unless there happens to be a defense attorney with a very good basis for deciding which of his cases may best be served by hiring a forensic scientist. No expert should bend the facts to meet the situation. The facts are the facts, and must fall where they may.

The same is true for prosecution experts. They are made available as employees of the state that may help the trier of fact to understand the situation as it applies in a particular case. The best scientists are advocates for their cause in that they are intimately involved in the state's maintenance and upkeep of the equipment used in prosecuting these cases, but they should be open minded enough to realize there may be two or more sides to each situation. Like the defense experts, they should not be present as an advocate for the prosecution, but only as an advocate for the truth and assisting the trier of fact in understanding the issues to make the best decision for everyone involved. Occasionally, one expert or the other will be used not as a scientist to explain the facts, but only as a rebuttal witness to present other possible explanations than those opinions presented by the opposing expert.

Frequently, there is a certain level of animosity on behalf of the prosecution utilized state employed scientists towards the defense experts. They try to portray the defense experts as "hired guns" out to make a buck. Sometimes, defense experts have been made to look like they are prostituting their trade for the defense. While working as a consultant to the defense in the past, the author has been asked by prosecutors and judges alike if he has ever done any work for the state. The answer should be obvious since the state has its own experts already on payroll. If would be incredibly foolish for the prosecutor to hire an outside expert consultant at considerable expense when there are already scientist with the appropriate expertise working for and available to the state.

The Judge or the jury has the obvious duty to listen to the evidence and apply the appropriate weight to the testimony given. If the prosecutor and the defense attorney have the basic knowledge and understanding of the scientific issues, and have presented their case properly, then the Judge or jury should have the information necessary to make the best decision in the situation at hand. The more complete the understanding on behalf of the trial attorneys, the more clearly the information should be when the decision of guilt or innocence must be made. The information needs to be presented in a manner that will be easy to understand and should not leave any questions unanswered in the minds of the decision makers.

There is another segment of the population with an interest in the outcome of the trials in our courts today. It is not uncommon when sitting in court watching these proceedings, to see people sitting in the courtroom with clipboards and taking notes. These are frequently members of MADD and/or SADD, and they attempt, sometimes subconsciously, to pressure judges and prosecutors to try and convict all suspects charged with DUI. There have been incidents where courts have been chastised publicly in newspaper articles or letters to the editor for being "soft" on DUI suspects. Many times, these people may not understand the issues involved, and why it may have been appropriate that an acquittal or reduction of charges may have been handed down. This certainly places undue stress and pressure on some courts to place less than proper weight on some testimony, or even totally ignore scientific evidence in some cases. Some of these public activist groups, even though acting on the best of motives, are acting more on emotions and less on facts. They cannot, or will not, understand the very complex issues involved in making the decision in light of all of the evidence as presented in a particular case.

One particular case in which local MADD representatives became involved was one that involved a fatality. The family of the victim (the driver of a vehicle struck head on by another vehicle) made significant attacks before and after the trial of the offending driver. What was sadly missing from all of the newspaper articles and letters was the fact that the victim had a higher BAC than the defendant. This isn't to say that the defendant should not have been driving, or that he didn't deserve to be tried for DUI, but facts of the accident indicated that the victim was impaired well beyond the level of the defendant. The coroner's report of autopsy indicated an extremely high BAC. What never came out in the media was that the victim may have not only contributed to the accident, but may have been able to avoid it had he not been driving under the influence. The decision made by the court was attacked by the media for weeks after the trial, although the court made an appropriate decision and disposition based solely on the evidence presented by both sides at trial. What looked superficially like a bad, or unjust, decision was in fact a decision the court agonized over but was necessary and based on the facts of the case.

This text has been developed with several purposes in mind. The crime of driving under the influence of alcohol is an offense deeply based in science, from the initial point of detection, to case development, right up to testing. There is little information available in an all inclusive work to help the key players to understand the basic scientific premises involved. This an attempt to remedy this fact, and assist in the ability of anyone to understand this complex scientific issue more fully. As an expert on alcohol physiology, the author is frequently asked for private lessons about some issues involving alcohol and the law. This is a direct attempt to provide a place for these answers to be available and accessible on a case by case basis. If this information is approached with an open mind, and is studied with an open mind, it may very well make developing cases and prosecuting or defending them a much easier prospect. It should be clearly noted that there are more issues than may be apparent from a detection standpoint, from a field evaluation standpoint, and from a testing standpoint than has been instructed in the past in current and accepted manuals and texts. Every effort is being made here to be fair from all avenues and approaches. We must remember that DUI is a criminal offense, and carries the same burden on behalf of the state as any other criminal offense—the defendant is innocent until proven guilty beyond a reasonable doubt.

Hopefully this work will become a useful reference tool for anyone with an interest in the scientific issues involved in DUI, whether from a defense, prosecution, or even public point of view. It is important that the social drinker wishing to be safe from injury as well as prosecution understand the limitations in consuming alcohol and later operating a motor vehicle. This is an easier concept for one that understands the metabolism of alcohol. This could help the responsible drinker remain within the laws and public expectations safely. The author has had many inquiries over the years from social drinkers seeking information about drinking and remaining safe without becoming a public danger. The best recommendation would be to not drive if you have had anything to drink, but there are certainly times when some alcohol consumption may not be an issue due to the physiology of alcohol metabolism.

Defense attorneys could benefit from a more thorough understanding of the scientific issues in evaluating how best to defend a client, or even if there is merit to aggressively put forth a defense at all. Sometimes the best defense is facing a conviction head on and limiting a client's exposure to the system as much as possible to gain the best outcome for all involved. In order to maintain appropriate standing with a particular prosecutor or court, the well informed and educated defense attorney will be very reluctant to put on a frivolous defense with no merit, yet when the defense is available and appropriate, can put on a strong defense grounded in solid scientific principles.

Law enforcement officers and prosecutors that understand the sciences behind the law and the offense more fully can better judge which case may be best for forceful prosecution, and which are better pled to lesser offenses in the interest of doling out appropriate justice. From the prosecutorial standpoint as well as the defense, it may not be in the best interest of justice to obtain a conviction at all cost, or an acquittal at all costs on behalf of the defense. The goal of both sides of this issue should be appropriate justice. The fairest and most professional prosecutors, officers and defense attorneys are interested in ensuring justice at all costs, certainly not just a conviction or acquittal.

There are volumes of research on DUI and other alcohol related issues available for review. It is inconceivable for any one person to spend the necessary time reviewing these articles and gleaning the important issues from each. Through this venue, an attempt will be made to incorporate much of the factual research available in the scientific literature. By placing this information in a single location, anyone with an interest in the science of alcohol and impairment may be able to have their questions addressed more directly. Many times different research articles are referenced in court by opposing counsel questioning experts, and it becomes apparent that there may be some fine point of the research being missed. It is not enough to read the conclusion and assume that single paragraph tells the entire story. Many times, one research article may seem to contradict another. Does that mean that one is incorrect? Certainly not, as not every research protocol is designed to produce the same result. What may look to be inconsistencies may simply be a difference in protocol that might indicate a simple explanation of the apparent contradiction.

Research into the standardized field sobriety tests has been ongoing for the past thirty five plus years, and has been updated and republished even as recently as the past several years. There are what appear to be differences in the conclusions drawn during the earlier and the most recent studies. With a better understanding of these tests themselves, and the full background of the studies, it is easily shown that the studies have very similar conclusions, and even can be used to prove premises other than those the researches ever intended.

Alcohol testing is an ever changing and evolving process that demands anyone with an interest in this topic aggressively maintains current knowledge. If not, there may well be an inability to correctly understand the results and the meaning to the outcome of a trial. In a society of laws that demands proof of guilt beyond a reasonable doubt, it is extremely important to know what the result indicated by the evidential testing device actually shows. Even if the legislature of a state dictates that the result produced by an instrument is definitive, it cannot legislate the science involved in getting to that point. The laws of science and nature are not similar to the laws of man, and cannot be changed by the stroke of a pen. Each of these scientific laws must be applied as they are set out. There is no bending of these laws, as each are laws of nature, and cannot be changed by one who doesn't like the meaning of a particular law. Each law of nature may have certain facts that must be present, and the outcome may only change depending on changing certain factors. If a gas law states the pressure is directly related to the temperature, then that will always be true. A change in one fact will provoke a change in the other with predictable results. Laws of determining blood alcohol levels written by man must conform to the laws of natural science, or they will always be open to argument.

Chapter Two

Alcohol Physiology

Pharmacology

Alcohol is a term that identifies an entire class of chemicals, not simply a single agent that we generally think of when we hear the term. Alcohols in general are a chemical identified by its chemical composition as well as its activity and behavior. Any chemical with an oxygen and hydrogen combined together (a "hydroxyl group") bound to a carbon atom is chemically considered an alcohol. These chemicals as a group are relatively polar (they have a positive and negative end, similar to a battery), which means they dissolve freely in water, and are used as biological solvents and carriers for other chemicals. All alcohols are biologically active with the potential for intoxication and all can produce fatalities from abuse and overuse.

The alcohol to which we refer in a legal sense when discussing impairment issues is actually ethanol, which is the alcohol involved in all intoxicating beverages. It has many other uses in daily life, and therefore has some situations where one may encounter this chemical other than as a beverage. This is the alcohol in "gasohol" where it is used as a gas additive and extender. It is used frequently as paint thinner and cleaner, and is used in many medications to dissolve solid chemicals and to enhance the effectiveness of some drugs. Ethanol is generally colorless and appears clear unless mixed with other liquids. It is odorless for the most part, although there may be some odor detected when concentrated vapors are in a closed container.

Ethanol is the alcohol responsible for the intoxicating properties of alcoholic beverages whether those beverages are beer, wine, liquor or other spirits. This is the chemical that provides the buzz or euphoric high associated with drinking that is sought after by those consuming alcoholic drinks.

There are other alcohols with common usages in daily activities, but none are used in a situation where they will be consumed in any amount to cause a measurable level on a regular basis. There may be some situations with these other alcohols where there may be accidental consumption, or even absorption in an industrial situation, and may cause some measurable blood level, but most are extremely toxic in any amount that may be measurable. Ethanol is the single alcohol most likely to be encountered in any amount to cause a measurable blood level and be involved in any medicolegal content.

Ethanol is the most widely used and abused drug in the world today. It has become so commonplace in today's society that it is rare to encounter anyone not touched by the effects of ethanol in some manner. Ethanol is used not only in alcoholic beverages, but also as a solvent in many medications, both prescription and over the counter. With the ability to dissolve many chemicals that normally won't dissolve in water, it is a necessary vehicle in order to make many medications into a liquid form. There are many topical creams and ointments that have the chemical imbedded into the base using ethanol to deliver the agent in a usable form. It actually takes considerable effort on the part of a patient to obtain a liquid medication that is alcohol free.

Previously, alcohol was utilized as a medical agent, either to provide anesthesia for dental or surgical procedures, or as a topical agent to sterilize the skin before a surgical procedure or to clean a wound as part of the treatment process. When used as an anesthetic, it would be provided to the patient in an amount sufficient to deaden the pain receptors and provide some level of comfort. A little more ethanol and the level of consciousness could be altered in order to provide more of a general type of anesthesia. It was recognized as a less than ideal anesthetic agent due to the extremely close levels required for general unconscious sedation and coma or death. When a little whiskey provided pain relief, a little more allowed surgery at a semi-comfortable level, and yet a little more caused the demise of the patient. It became clear that newer and better tolerated agents needed to be developed for medical anesthetic purposes.

The polarity of the ethanol molecule provides the properties that cause the rapid absorption and dissemination of the alcohol throughout the body. Due to the fact that the molecule mixes quickly with the water in the body and the tissues, it passes through the bowel wall and into the circulatory system quickly and without the need for actively pumping the alcohol from one area to the other. Once into the blood, it quickly spreads throughout all of the body's water with which it comes into contact. Since the largest portion of our bodies is water, the ethanol consumed is quickly diluted into all of the water spaces of the body. The limiting factors are the tissues through which the alcohol must pass in order to be absorbed, and the barriers from one tissue type to another. Impairment occurs when the ethanol reaches the cells of the central nervous system. Ethanol in the mouth, stomach, bowel, and blood cannot cause any level of impairment; it must reach the brain in order to cause any measurable level of intoxication and impairment.

Because the ethanol reaches all body tissues relatively quickly once absorbed, it must be understood that the process of elimination begins immediately once the ethanol is absorbed into the blood. It reaches the sites of elimination just as quickly as it reaches the sites that cause impairment. Ethanol is eliminated by several different mechanisms, and most are dependant on the concentration of alcohol in the blood that reaches these sites for elimination. Blood alcohol levels are an orchestrated balancing act between absorption and elimination. One can take steps to control the rate of absorption of alcohol from the gut into the blood, but the elimination rate, once it reaches the maximum point, is very difficult to manipulate. If the absorption is held to match the physiological elimination rate, blood alcohol levels will naturally remain at or near zero. It is important to understand that elimination begins immediately upon absorption. Only when absorption rates surpass the maximum elimination rate can the blood alcohol level rise above zero.

Absorption of Ethanol

Absorption of alcohol begins immediately with consumption. When any alcoholic beverage is consumed, it passes from the mouth to the esophagus, past the lower esophageal sphincter, and into the stomach. The mucosal surfaces of these areas are such that there is very little absorption that takes place in each of these areas as they are designed to be a barrier against absorption of consumed foods and beverages in order to allow the stomach to begin the process of digestion. Anything consumed and absorbed before the stomach will not benefit from the digestive process as begun in the stomach. Although some alcohol can enter into the blood stream through these layers, it is of such a small amount that it is generally not of any measurable consequence.

From the lips to the lower esophagus, the membranes are architecturally not designed to provide for much more than transportation of consumed items into the stomach. The stomach itself is essentially a holding tank with the secondary function of mixing the consumed food and drink with hydrochloric acid and several other lesser digestive enzymes to begin the process of digestion. Again, there may be some level of absorption of a chemical as polar as ethanol, but it is generally not of any significant level to be considered in the grand picture of alcohol absorption and the ensuing blood alcohol curves.

Alcohol is absorbed by simple osmotic diffusion across the membranes of the small bowel. Once alcohol enters the small bowel, it passes through the membranes into the circulation of the fine network of capillaries designed to move nutrients from the gut into the blood system. Some nutrients require energy to force the molecules from the inner aspect of the bowel into the deeper tissues to be absorbed, but with ethanol, the molecule is so small, and so polar, that it just moves by simple diffusion across the tissues and blood vessel walls. There is no chemical "pump" needed as is required in some larger molecule nutrients. The only requirement is that the concentration of the alcohol in the contents of the bowel must exceed the concentration of the alcohol in the blood. The passive diffusion of alcohol from one area to the other is dependant on nothing more than a concentration gradient to push the alcohol form one area to another.

Many factors control how quickly the alcohol consumed is absorbed into the blood. Before any major amount of alcohol can be absorbed, it must get into the small bowel. At the lower end of the stomach is a valve that controls the flow of stomach contents into the small bowel. The pyloric sphincter functions to control the flow of stomach contents into the duodenum (the first segment of the small bowel) where absorption begins at any significant measurable level.

Food present in the stomach can alter the pressure of the pyloric sphincter and the amount of stomach contents permitted to move into the small bowel. Most foods are complex mixtures of lipids, carbohydrates, and proteins. High fat (lipid) content foods tend to slow the release of food from the stomach due to the fact that the body will attempt to hydrolyze the fats before presentation to the bowel. Foods with high carbohydrate (or complex sugar) content tend to speed the release of stomach contents to the bowel. Proteins tend to have less of an effect to the positive or negative on presentation of foods from the stomach to the small bowel.

It is easily seen that anything that slows the movement of food from the stomach and into the small bowel will certainly slow the presentation of alcohol mixed with the stomach contents to the small bowel. The inverse is also true. Anything that may speed the presentation of the stomach contents to the small bowel will increase the presentation of alcohol within the stomach contents to the small bowel. The other issue that must be understood with food is the dilutional effect of the food to the alcohol. If the alcohol is mixed with a significant amount of food of any type, or other drinks, then the concentration within the stomach is much less than that when the alcohol is mixed and consumed.

Timing of foods consumed is as important as the types and amounts of foods consumed. If the food is taken along with the alcoholic beverage, it is going to be present at the same time and will affect the movement of alcohol from the stomach and into the bowel. If the alcohol is consumed after a meal, depending on how much time elapsed from the meal to the drinking, the food may have already moved from the stomach and will have much less impact on the movement of alcohol into the small bowel.

Much has been made over the years about alcohol consumed on an "empty" stomach. The general conception is that any time one is to drink on an empty stomach the alcohol will be quickly and completely absorbed with corresponding rapid elevation of blood alcohol levels and therefore impairment. This may or may not be true as there are other variables that must be considered. Concentration of the consumed alcohol can have several effects on the ability of the body to absorb alcohol.

There is an ideal recognized concentration of alcoholic beverages for absorption. Any alcohol that is hyperconcentrated, or too strong, may significantly slow the absorption of the alcohol by causing the pyloric sphincter to close and delay the presentation of alcohol to the bowel. This is a defense mechanism whereby the body will attempt to protect itself by preventing the movement of a strong concentration of alcohol to absorption sites. Alcohol will be slowly released through the pylorus to the small bowel. This phenomenon is ignored by many professionals in the study of blood alcohol curves, yet it has been proven over and over again.

Any time the alcohol solution is very weak there will be slower than expected absorption due to the fact that less alcohol is available to the blood supply of the bowel. If the ideal concentration (generally thought to be about five to twelve percent) is presented to the digestive tract, then the maximal absorption level can be expected. This ideal concentration just happens to be the same as that seen in most beers and wines sold commercially. Beer is slightly more quickly absorbed, due to the carbonation. The dissolved gases in the beverage aid in absorption to a small degree.

Normal circadian rhythms also affect the speed with which alcohol may be absorbed. These rhythms are normal biological changes that take effect in the system depending on time of day. All biological systems are controlled by these naturally occurring rhythms. Any alcohol consumed earlier in the day tends to be absorbed more slowly than the alcohol consumed later in the day near the end of a long cycle of wakefulness. This may be affected by hormonal influences, as well as basal body temperatures. As the biological system is further from an overnight sleep state, there will be higher basal metabolic temperatures, and higher need for caloric intake to feed the higher metabolic needs. These factors could influence the absorption rate permitted by the gastrointestinal system.

Elimination of Ethanol

It has already been mentioned that elimination of alcohol begins immediately upon absorption. The blood alcohol level is always a balancing act between absorption rates and elimination rates. So long as there is more alcohol being absorbed than eliminated, then the blood alcohol level will rise. Immediately upon complete absorption, then elimination rates will overcome that of the absorption, and the blood alcohol level will begin to plateau and then drop off at a predictable rate.

Alcohol consumed and passed into the small bowel is absorbed into the circulation via the capillaries of the portal venous system surrounding the small bowel. These small vessels coalesce into large veins, eventually joining into the portal vein, and provide transport directly to the liver. At this point, the liver detoxifies a significant amount of the alcohol consumed and absorbed into the blood. It is estimated that seventy five to ninety percent of all alcohol consumed and absorbed is eliminated by this first pass metabolism via liver enzymatic processes. This occurs before any alcohol reaches the general circulation, and therefore, the central nervous system.

As could be predicted, any condition that inhibits the function of the liver may alter the first pass metabolism. Certain disease processes such as viral conditions (hepatitis), chronic liver damage (from chemical exposures such as alcohol and environmental toxins), or effects of medications (Tylenol and many heartburn medications) may drastically alter the first pass metabolism enzymes to such a degree that alcohol levels could rise significantly higher than expected.

What ever alcohol survives the first pass metabolism moves beyond the liver, and rapidly diffuses throughout the body water as the blood containing the alcohol circulates through the body. Being very polar and very water soluble, all of the watery tissues of the body are quickly brought to equilibrium with the blood's alcohol concentration. As alcohol reaches all of the tissues of the body, it also reaches all of the major sites of elimination.

As the blood circulates, it returns over and over again back to the liver. As the main chemical detoxification organ of the human body, the chemical processes of the liver are brought to bear on all of the alcohol in the blood with each pass through that organ. It is an extremely vascular organ, with an incredible ability to detoxify an amazing amount of offending substances. This is an extremely complex mechanism and is chemically mediated by more than simple filtration.

All of the circulating blood also passes through the kidneys. The kidneys are the second most important organs of alcohol elimination. The process of the kidneys removing alcohol is that of simple filtration and osmotic diffusion. As the kidneys filter the blood to remove all the toxic, unnecessary components, the water that leaves the kidneys also carries a corresponding level of alcohol from the body. This is the basis of urine testing to determine blood alcohol levels, as the urine will have the same alcohol concentration as the blood that is being filtered. On the surface, it may appear that urine testing would be a reliable manner of measuring blood alcohol concentration, but this in itself has inherent dangers that have made it unacceptable in most jurisdictions. This issue will be dealt with in another portion of this text.

Blood carries alcohol to the subcutaneous tissues that have the sweat glands imbedded within them. Any perspiration produced will also carry with it a corresponding amount of alcohol of the same concentration as the blood at the time of the sweat production. Naturally, the amount of alcohol eliminated in the sweat will be minimal and will be variable depending on the ambient temperature, possible illness (fever), and nervousness, all conditions that may increase sweat production.

Blood presented to the alveolar surfaces in the lungs will also pass alcohol vapors into the air inspired into the lungs just as it does carbon dioxide. This process is also concentration dependant and follows the natural gas laws of physics. Inspired air generally has less carbon dioxide and ethanol vapor, and more oxygen than the blood surrounding the alveolar sacs, and this is what allows oxygen to enter the blood, and carbon dioxide and ethanol to leave the blood. The very complex gas laws will cause a predictable amount of ethanol to be in the breath, and thus we have the basis for breath testing for determining blood alcohol levels.

Some alcohol will pass unchanged and unabsorbed into the large bowel, and may leave the body in the feces. It is easy to understand that any time the stools are watery then there will be more alcohol leaving the body by this mechanism. Anything that increases the transit rate of the bowel content to the rectum will increase the amount of alcohol that may leave the body by this route. In addition to the alcohol that passes through the bowel, if there is considerable water content to the stools, and the blood has a higher concentration of alcohol than that watery stool, then there may be a movement of alcohol back from the blood into the bowel, exactly in the same manner as absorption, but in the opposite direction.

Each and every cell in the body has the ability to burn alcohol for fuel, and some of the absorbed alcohol molecules will be consumed by the metabolism of alcohol as fuel in each of these cells. These molecules are burned and converted to water and carbon dioxide which is returned to the blood for elimination elsewhere.

All of these mechanisms come into play additively to the give the elimination rate of the alcohol from the body. The major mechanisms are the liver and kidneys, as all the other mechanisms are small players compared to these two organs. Any attempt to predict the elimination rates of alcohol must take into consideration the health of these two organ systems, and the presence of any potential interfering substances that may alter the functions of these two organs.

Some medications may significantly alter the function of elimination. If a subject is on a diuretic, and has increased urine output, then naturally more alcohol will follow this water through the kidneys and the rate may be increased. Sugar is a natural diuretic, and will act in the same manner as a chemical medicinal diuretic. Medications that are used to treat peptic ulcer disease (H-2 blockers such as Tagamet, Zantac, or Pepcid) or gastroesophageal reflux disease (proton pump inhibitors such as Prilosec, Nexium, Prevacid, Protonix, and Aciphex) can cripple the liver enzymes that are used to detoxify the alcohol molecules in the liver, and the alcohol levels can rise quicker and remain higher for longer periods than expected. Over the counter drugs as common as acetaminophen (Tylenol) also affect liver function and can inhibit the elimination of alcohol. All of these drugs are readily available both as prescription and over the counter, and are some of the most commonly taken medications today.

On the opposite side of the issue, elimination rates generally tend to be higher in subjects that drink alcohol on a regular daily basis. These subjects tend to have higher than expected elimination rates due to the fact that their liver function and enzymatic processes are enhanced by the frequent presence of alcohol and the need to be able to be more efficient at its removal. This is a chemically trained response, and needs to be taken into account when attempting to determine alcohol elimination curves. It involves no effort on behalf of the subject, and is automatic in development of a protective process in the body.

Normal elimination rates can be predicted at approximately 0.016% to 0.018% per hour. In heavy drinkers and alcoholics, this rate may rise to as high as 0.020% to 0.026% per hour. Differences in elimination rates between men and women tend to be more body fat related, temperature related, and hormonally related, but can still be factored into attempting to predict rates.

It is important to understand that absorption and elimination rates are very difficult to pinpoint. This fact is often used to argue against plotting any type of alcohol curve for predicting alcohol levels over time. The inability to pinpoint the rates exactly does not prohibit some degree of accuracy in plotting an alcohol versus time curve, if all the necessary factors are taken into consideration. It must be understood that there are some variations in absorption and elimination rates from one person to another, and even within the same person from day to day. It is important to state estimated blood alcohol levels as a potential range depending on the factors involved, and taking into consideration all of the necessary factors.

So long as it is understood that blood alcohol levels are a balancing act between absorption and elimination, and elimination begins immediately upon absorption, then there can be some degree of accuracy in predicting blood alcohol levels. It is true that there are many factors involved, and some forensic scientists will use these different factors to argue that predictions should never be attempted. Care must certainly be taken, but there is some reliability to comparing test results with predicted results. Frequently, there will be good correlation when all of the factors are utilized, and a thorough and accurate medical history, drinking history, and social history are obtained.

Alcohol Impairment

The important issue in dealing with alcohol and operation of motor vehicles is impairment. Alcohol is an impairing compound, and taken in sufficient doses, it tends to cause impairment up to and including coma and death. It is necessary to enter the central nervous system in order to cause any impairment. All impairing effects take place in the brain, and no where else. Alcohol must get from the bottle to the GI system, to the blood, and then to the brain to cause impairment. This is what makes blood alcohol levels such an important issue when dealing with DUI and driving impaired.

As alcohol levels rise in the arterial circulatory system with absorption, it approaches a level that overcomes the blood/brain barrier and the alcohol begins to enter the central nervous system. As the alcohol level begins to rise within the central nervous system, the alcohol begins to interfere with the transmission of signals from one cell to another, and cellular communication begins to be disrupted. The effect of cellular communication disruption is dose dependant, and as the alcohol dose increases, the level of interference also increases. This is observed externally as signs of impairment, with what we normally see as intoxication. This appears in stages depending not on the amount of alcohol consumed, but on the amount of alcohol that has reached the central nervous system. There have been many attempts over past decades to grade levels of intoxication and impairment in individuals, with little success, as there are so many factors to apparent impairment.

Outward signs of impairment are essentially the same for everyone that consumes alcohol, except that the measurable blood levels may be significantly different from one person to another. Initial signs of impairment, although similar in appearance, will begin to be noticeable at different levels in different people depending on drinking experience, body fluid levels and other physiological factors, medications that may be on board, and levels of fatigue. If one were to alter any of these variables from one day to the next, then levels of impairment may be noticeable at different blood alcohol levels for even the same person from one day to the next. Every state has a level of presumption for impairment, and to reach or exceed that level is all that is necessary to prove legal impairment regardless of actual impairment. At any level below the presumptive legislated limit a subject may be considered impaired in light of other evidence of alcohol effects.

The earliest signs of impairment would be the loss of inhibitions. This is normally the entire purpose of consuming alcoholic beverages at the outset. It is normally consumed in social setting as a relaxant and to allow one to be "sociable" by removing a subject's normal level of inhibition. One would become more talkative, more likely to interact on a social basis (dancing, singing, or showing a higher level of self confidence, even when not warranted), and possibly more argumentative. This level of impairment is generally seen in the range of 0.02% to 0.05% depending on alcohol tolerance.

The next signs of impairment would be giddiness. A subject would become happy and carefree, sometimes to the point of annoyance. They will be the "life of the party", telling jokes, acting sometimes aggressively social, and possibly bordering on irritatingly pleasant. This occurs at just slightly higher blood levels than seen for the loss of inhibition stage, sometimes even without the consumption of more alcohol, but just simply the absorption of what was already consumed.

As blood alcohol levels continue to rise to the range of 0.04% to 0.08% we begin to see mild to moderate sensory losses. There may be some visual acuity loss, and some peripheral visual disturbances. With this peripheral visual disturbance ("tunnel vision") a subject may be noticed turning his head side to side to look at people or things not directly to their front. As alcohol levels continue to rise, one may be noticed to start turning their entire bodies toward members of a group in a conversation instead of just turning their heads. Color vision will begin to become altered. Fine sensory deficits may begin to appear, and subjects may have to be more cautions when picking up items from a table or the floor, many times appearing to the sober observer as being slow and exaggerated in digital manipulation.

Next steps noticed in impairment would be some mild to moderate behavioral changes as decision making skills may begin to be impaired. A person in this stage may begin to act out inappropriately, comments may seem out of context with a conversation, and more significant fine motor manipulation will be obvious. Eye and hand coordination will begin to suffer at this stage. There will be apparent difficulties with performing tasks with multiple stages or steps. This level of impairment is noticed at the 0.08% to the 0.12% blood alcohol range.

From 0.10% to 0.15% blood alcohol levels gross motor impairment begins to become obvious. Subjects begin to have exaggerated movements involving walking and turning, as well as difficulty rising from a sitting position. They may appear to use extra caution when going from standing to sitting in a chair, appearing to have difficulty ensuring they are going to hit the seat as they attempt to sit. When walking among guests in a crowded room, they may have to put a hand on each person they pass in order to prevent "bumping" them as they walk. When eating, there may be some slowness and deliberate caution getting food onto utensils and into the mouth.

As a subject advances in blood alcohol levels at and above the gross motor skill impairment levels, they begin to appear to be openly drunk. This is clearly beyond what would be considered merely impaired, and is a significant advance in outward signs of drunkenness. As someone begins to approach the 0.16% to 0.18%, they begin to exhibit signs of altered consciousness. They begin to appear drowsy or sleepy, and may actually nod off if left alone not engaged in an activity or conversation. When engaged in conversation, they answer very slowly as they must think about the question and process it in an alcohol altered brain before attempting to formulate a response.

When more and more alcohol is consumed and absorbed and blood alcohol levels begin to rise above the 0.18% level to the area of 0.24% most people will show very clear signs of severe intoxication. They will exhibit clear signs of impaired consciousness with stupor and inability to perform any dexterity or motor skill tasks of any type. These subjects will be obviously drunk, and unable to perform even gross motor tasks. They will be difficult to reason with, as even the reasoning centers in the central nervous system will be deeply impaired.

Above these blood alcohol levels, beyond the range of 0.25%, most subjects will begin to enter the range of alcohol induced coma. The apparent drunkenness is clear, and the subject will be lethargic at best. In the range of 0.30% to 0.35% nearly everyone will be in an unconscious state, or on the verge of actual coma. As blood alcohol levels reach the area of 0.40%, death from acute alcohol intoxication is a genuine concern, and it is rare that anyone would be operating a motor vehicle at these elevated levels.

These levels as described here are simply nothing more than general guidelines, and most of these ranges are variable from one person to another and even within the same person on different days. There is always going to be the subject that has been documented as being much more intoxicated by blood alcohol measurements, yet is still more functional than these guidelines may indicate. This is not uncommon, and care should be taken when using any set of guidelines to indicate certain expected impairment observations for certain blood alcohol levels. There has actually been credible scientific and medical documentation of subjects in emergency rooms acting mildly intoxicated, yet producing measured blood alcohol levels in the 0.40 plus areas. These subjects sometimes develop more intense signs and symptoms of intoxication as they are observed in the emergency department, but tend to surprise everyone with their ability to function and answer questions appropriately, even at these significantly elevated blood alcohol levels.

There are several types of alcohol tolerance that can contribute to a person's ability to function well outside the normal ranges of expected impairment at higher blood alcohol levels. The existence of alcohol tolerance is occasionally denied by some forensic scientists that deal in this area, yet it is very well documented and demonstrated by many different researchers over a great number of years. It is highly irresponsible from a scientific standpoint to deny the effects of tolerance to alcohol impairment in any individual.

Alcohol tolerance can be most easily demonstrated by comparing three different types of drinkers. The first is the novice drinker, a subject that has little or even no experience with alcohol consumption and its effects. With only one or two drinks, this subject may reach a blood alcohol level of 0.03% to 0.04% and be sloppy drunk by appearance of observable impairment signs. They could even be close to having altered levels of consciousness from alcohol consumption, all at levels that should not even be noticeable in a normal adult social drinker. The second drinker is this normal "social" drinker that consumes alcohol on a somewhat regular basis, but not on a day to day pattern. This person may have several drinks in a social setting several times a month, or even one or two drinks or glasses of wine with dinner several times a week. This person will normally follow the guidelines listed for impairment levels stated earlier in this section. If they occasionally have the opportunity to consume a significantly larger amount of alcohol, they could appear quite impaired at the levels expected as explained above.

The third subject is the daily alcoholic drinker. This person may awaken each morning with a measurable amount of alcohol already present from the consumption of the day prior. The drinking for the new day begins at breakfast with vodka in the coffee or the orange juice for breakfast. This person continues to drink during their morning breaks even while working, and then tends to have a "liquid" lunch. The afternoon may continue the same as the morning with several drinks over the afternoon, and then for dinner, there is the mandatory bottle of wine.

With this type of drinker, they may appear perfectly normal with blood alcohol levels in the 0.20% to 0.25% range. Their driving may appear normal and if stopped for some minor traffic infraction, and the odor gives away the presence of alcohol, they may do quite well on field sobriety tests. A law enforcement officer may be quite surprised to see a very high reading on a preliminary breath test, and may even question the accuracy of his instrument.

There are two mechanisms in this scenario that are occurring that would allow for the amazingly high blood alcohol level and apparent lack of impairment. The first is behavioral tolerance and the second is chemical tolerance.

With the behavioral tolerance, the subject is so used to the high level of blood alcohol that he actually trains himself to perform quite well even when intoxicated. Repetition is the most normal and common form of training in any animal species, including human animals, and sheer repetition of normal daily function while intoxicated will allow one to perform quite normally when blood alcohol levels are surprisingly high. The old adage of "practice makes perfect" is the best way to describe this mechanism, and is easily understood in law enforcement circles by understanding normal police training procedures. Officers are trained in firearms on what are known as "combat courses" that simulate firing under stress and quick decision scenarios that develop "muscle memory". The same is true about self defense tactics. By sheer repetition in training, officers hopefully will respond almost as second nature when lives, their own or those of the public, are at risk. This is a behavioral modification, very similar to that of the alcoholic that has learned to function with a high blood alcohol level with little or no sign of impairment.

The chemical tolerance is the body's ability to protect itself from the constant presence of the alcohol as a toxin. It develops biochemical capabilities to handle the increased levels of alcohol present on a regular daily basis, and allows the central nervous system to function in what may appear to be a near normal manner. This occurs by forming new neural pathways, as well as significantly enhanced enzymatic processes. This is the same mechanism that increases the elimination rate in heavy drinkers above that expected for most social drinkers. The biochemical pathways are honed and enhanced to handle the higher, constant levels of alcohol as a self preservation tool by the body of the heavy drinker.

A regular heavy drinker that abstains for one reason or another, either by choice or by force, and has a blood alcohol level that falls off to zero may be more of a danger than when they actually have some alcohol present. This heavy drinker will become shaky, unsteady, tremulous, and physically ill when alcohol is suddenly removed. There may be significant apparent levels of impairment, and even altered levels of consciousness as blood alcohol levels approach 0.00%, exactly the opposite of what would normally be expected. This situation will begin to be noticed as blood alcohol levels begin to drop below those normally experienced by a subject, and is variable depending on the amount of alcohol normally consumed by that individual. This is why most alcoholics need to have that first morning drink, like most people need their first cup of coffee in the morning. Without that first drink to boost the blood alcohol level, a subject may appear quite abnormal.

This is not the same as the delirium tremens (DT's) seen in alcoholics that have their alcohol suddenly removed, as this will be the next step in this sequence. The DT's generally take 72 to 96 hours to appear, and are much more pronounced than just acute alcohol withdrawal symptoms. DT's are generally not seen in subjects outside a medical setting, as most legal situations generally would fall within the first 72 hours, and therefore would simply be acute alcohol withdrawal. When it does appear in a law enforcement setting, it is generally in an incarceration situation, and not an actual arrest that causes an encounter that may last only several hours. Alcohol withdrawal that results in DT's will be addressed later in this text as an issue that needs to be understood for safety issues of persons that may be in custody and to prevent civil liabilities for law enforcement personnel.

This should in no way be construed as any form of approval, or any form of medical or scientific recommendation, that any person should be permitted to operate a motor vehicle at blood alcohol levels above the legislated legal limits because they are "more safe" with some alcohol in their system than without. Even though the outward appearance will be that they are not impaired, there is still some level of impairment that absolutely should not allow this person to safely operate a motor vehicle, for their own benefit and safety as well as that of the general public. This should never be used as an argument that any subject was less dangerous or impaired due to the presence of alcohol.

Another form of tolerance is mild and noticed in nearly every single drinking episode of the minimal or social drinker. The signs of impairment are more pronounced early in the drinking episode each and every time. The obvious signs of impairment are present more as blood alcohol levels rise than when they peak, and even more so than when they are declining. It is not certain as to whether this situation is actually chemical, or simply a behavioral reaction to expectations of the effects of alcohol with that first drink of the evening. It is likely to be some of both of these effects, as the body's ability to handle the alcohol improve with time as the alcohol is metabolized.

At the opposite end of the spectrum is the subject that appears to be significantly more impaired in contrast to the measured blood alcohol level. As discussed earlier, this could simply be a novice drinker or it could be that there is some other type of intoxicant on board that is enhancing the impairing ability of the two intoxicants when mixed together. There are certain medications available, both by prescription and over the counter that can make one appear quite impaired, even without the coexistence of alcohol. Mix these medications with even a little bit of alcohol, and significant impairment may be present.

Antihistamine medications, such as Benadryl, diphenhydramine, Claritin, Atarax, hydroxyzine, and chlorpheniramine should always contain some type of warning about not driving or operating any dangerous equipment while using these medications. None of these products should ever be taken if there is any alcohol present in the system. Many of these medications have now become over the counter and are available to patients without benefit of medical advice, but are clearly marked on the package as sedating. Other medications with similar actions are still available by prescription only, and patients should be warned when given the prescription to not operate a motor vehicle while taking the medication.

Prescription pain medications in the narcotic classes include codeine, hydrocodone, oxycodone, morphine, hydromorphone, and other very similar types of drugs all are significantly impairing when used by themselves, and anyone on these types of medications should not drive until a level of tolerance is achieved that does not cause any altered levels of consciousness. Most health care providers would warn patients to not drive while using these medications even after tolerance is developed, yet FDA and manufacturers recommend care when driving, and only after seeing how these medications affect someone after time. ***At no time should any of these narcotics ever be mixed with alcohol in any amount***. There should never be any exception to this rule, as the effects of alcohol and these narcotics are exponential, and not simply multiplicative.

There are many over the counter sleep inducing medications, but these tend to actually be in the antihistamine class, and they harness the most common side effect of the drug (drowsiness) to induce sleep. Other sleeping medications, tranquilizers, and medications in the benzodiazepine class are available by prescription only and should never be mixed with any amount of alcohol. Several of these medications are frequently prescribed, and some are on the most common prescribed drug lists. Recent direct to consumer advertising by pharmaceutical companies has increased consumer demand and awareness of these medications, and they are becoming more and more visible in the market today. Lunesta, Ambien, Sonata, Valium and Xanax are some of the most commonly encountered medications in this group and all cause significant impairment in their own right by virtue of their desired effects. These are all very effective medications for their intended prescribed purpose, and because of their intended use, no one should ever operate a motor vehicle after using one of these medications. Any level of alcohol mixed with these drugs will cause significant impairment even with very low blood alcohol levels.

Any subject with a very low blood alcohol level but an apparent elevated level of impairment should be questioned about potential medications. If there is any question about impairment, and little or no alcohol is present either on preliminary breath tests or evidential breath tests, blood tests should always follow in order to determine the presence of one of these other drugs or the possibility of illicit street drugs.

Chapter Three

Determination of Impairment

Considering the most common alcohol related offense is driving under the influence, at some point it will be necessary for any law enforcement officer to determine if a driver may have had too much to drink to be operating a motor vehicle. There is currently a wide variety of tools available to the modern law enforcement officer today that aids the officer in this endeavor. There are flashlights with alcohol vapor sensing cells that will alert the officer to the presence of alcohol (or similar chemicals) in a vehicle or emanating from a subject being questioned. These flashlights have a simple light or other signal to alert the officer only to the presence of alcohol vapors. More elaborate tools, such as the field alcosensors available today can actually give a rough quantitative reading of breath alcohol levels. There is a standardized battery of field sobriety tests that have stood the test of time and multiple reviews for validity to allow the officer to actually make a determination as to when to arrest or not arrest so long as the test are applied appropriately. The ability exists with these tests to make this determination within a reasonable degree of scientific reliability due to the many years of research to develop this battery. Once the arrest is actually made, there are evidential breath tests available to actually quantify the breath alcohol level and provide certification for use in courtroom trials. On occasion, there may even be blood tests to measure blood alcohol levels, and the possibility of other impairing substances. It is important to note that the most accurate method of determining blood alcohol levels is this blood test—if one wants to know the blood alcohol level, then draw the blood and apply scientific principles to measure the amount of alcohol in the sample. This chapter that follows will not address these standardized field sobriety tests, as they are so complex they deserve a chapter of their very own.

Short of the officers own nose, or a flashlight sensor or similar device, the first tool likely to be used by law enforcement officers on the street is the field alcosensor. There are actually several different brands approved for use in many different states, and most specify more than one that may be utilized by officers in their own respective jurisdictions. Generally, field alcosensor test results are used only for probable cause to pursue an arrest, and are not admissible in court. In the Commonwealth of Virginia, officers are usually permitted to testify that the test was offered and accepted or declined by the defendant, although occasionally prosecutors may attempt to get the reading into evidence to "prove" if the blood alcohol level was rising or falling at the time of the offense and the test.

There are excellent forensic scientific reasons for not permitting the actual reading of a field alcosensor at trial other than to establish the fact that there was actually alcohol present at the time of the stop or accident. There are several guidelines that are used to determine the difference in an evidential test and a screening test. Any screening test should be inexpensive, readily available, and easy to use. There are none of the built in safeguards for accuracy and reliability with screening tests, as it is with the evidential breath tests. There are no regular simulator comparisons for ensuring accuracy, the instruments tend to be carried around and bounce around in the police cruiser (not mounted permanently as the evidential instrument is required to be), it is never used in a temperature and humidity controlled environment like the more accurate evidential breath test instruments, and there is no mechanism in place to shield the instrument from radio frequency interference, particularly in light of the fact that it may be used within inches of the officers radio transmitter.

The field alcosensors are not used with any effort to ensure a 20 minute waiting period to allow dissipation of exogenous mouth alcohol. These less expensive screening devices have no built in mechanism like the more elaborate evidential instruments for detecting this mouth alcohol. They may read other contaminates as ethanol, and give a significantly elevated reading. It is not unusual for an officer to obtain a significantly elevated reading on the field alcosensor, and 45 to 60 minutes obtain a much lower reading on an evidential breath test.

Experiments have been designed that show subjects with actual blood alcohol readings of 0.00% (*no* alcohol present) to give actual alcosensor readings as high as 0.14% after simply rinsing the mouth with 80 proof alcohol or mouthwash containing alcohols other than ethanol. The instruments tested in this manner are those used by most law enforcement agencies across the country, and are not simply cheap, knock off versions of testing devices. Clearly, since there was no alcohol swallowed, the readings were obtained only from residual alcohol in the oral cavity, and not by any alcohol in the blood or breath. Serial testing over 30 minute periods tended to show gradually decreasing levels of alcohol until all the alcohol in the oral cavity dissipated and a 0.00% reading was again obtained.

Increased Alcosensor readings after rinsing mouth with standard Mouthwash Solutuion

Time From Rinse with Mouthwash	Actual Alcosensor Reading
0 Minutes	0.00%
5 Minutes	0.142%
7.5 Minutes	0.122%
10 Minutes	0.068%
12.5 Minutes	0.021%
15 Minutes	0.00%

Chart A

The most commonly used evidential breath instrument in the Commonwealth of Virginia, and many other states within the United States, is the Intoxilyzer 5000. Virginia is currently looking at other instruments at the time of this writing, and may be making a change in the future. The last instrument used in Virginia for any significant period of time was the Smith and Wesson 900A. To quote a member of the Commonwealth of Virginia Department of Forensic Sciences staff, "...our current instruments are aging, and we are always looking at newer technology..." It can be assumed that there were no problems indicated with the current technology, but newer instruments were always worthy of evaluation.

The Intoxilyzer 5000 has programming that is designed to make it extremely reliable and accurate for legal blood alcohol determinations. The software to operate the computer aspect of the instrument is proprietary, and is closely guarded by the state. It is frequently insinuated that the software is not publicly disclosed, and therefore, no one can testify as to the operation of the particular variation of the instrument used in that jurisdiction. In fact, when the defense cross examines the state's scientist, they admitted that the "software" that is proprietary is the software that runs the printers, clocks, and other peripheral options of the instrument. The actual testing functions of the Intoxilyzer 5000 are set by the manufacturer, and this is the "guts" of the instrument that makes it what it is. This portion of the programming is considered "firmware" by the manufacturer and the state, and is not altered from one jurisdiction to another. Therefore, anyone with the knowledge to understand and operate an Intoxilyzer 5000 will have the basic knowledge to understand the operation of this instrument no matter what the proprietary software package may be.

The most troubling issue with the Intoxilyzer 5000 is that it measures breath alcohol, and utilizes certain assumptions (necessarily so) in order to convert this breath alcohol level to a usable blood alcohol level that makes sense for use in a trial to determine impairment. If one is to accept this instrument as totally accurate at measuring the actual breath alcohol concentration, and further accept all of the scientific assumptions built into this instrument, then there would be the reasonable expectation that the corresponding blood alcohol concentration reading produced by the instrument would be true.

There is a certain margin of error built into any scientific method of measuring anything. Unfortunately, absolute perfection is neither expected nor possible. CMI, the manufacturer of the Intoxilyzer 5000, states this instrument has a margin of error of only 10%, which is reasonable for this type of testing equipment. In general, this margin of error is important only when there is a borderline reading for determination of innocence or guilt, or there is a borderline reading at another cut off value, such as higher readings for advanced punishment and sentencing guidelines. When considering a reading of 0.08% (the current level of presumed guilt in Virginia), the 10%

margin of error may drop this reading to 0.072%, which is truncated to two decimal values in the Commonwealth of Virginia (which becomes a 0.07%). A reading of 0.20%, which would carry significant mandatory incarceration time, may potentially be 0.18% with the 10% margin of error, removing the subject from the elevated mandatory incarceration category to a lower punishment area.

In order for the Intoxilyzer to work at all, there are certain physiological assumptions built into the instrument in order to allow for conversion of the measured alcohol level in the breath sample to be delivered as a "blood alcohol" reading. The greatest assumption most commonly addressed is the partition ratio of 2100:1. This assumption is that for every unit of alcohol measured in the breath sample, there are 2100 of the same units in the blood of the subject that provided the breath sample. This 2100:1 ratio has been developed over time by comparing the amount of air in the alveolar spaces compared to blood in contact with these airspaces, as well as the temperature of the blood and air. These two physiologic values are important when making any attempt to compare the amount of alcohol that may move between the two fluids, blood and air.

Even though ethanol consumed as alcoholic beverages is a liquid, when attempting to understand the movement and equilibrium of alcohol between the blood and lung air it must be understood that ethanol is a gas, or vapor. The movement of the ethanol between blood and air is governed by the scientific ideal gas laws, and a basic knowledge of these gas laws will go a long way to assisting in understanding the functioning of the Intoxilyzer 5000 and the assumptions in producing a blood alcohol level.

Alcohol is a vapor dissolved in the blood when it is absorbed. It is distributed in the total body water, to include the blood, which of course is primarily water. Henry's Law of gas physics explains how any gas (vapor) dissolved in water will interact with the air above the surface of the water. Every chemical has its own affinity and ability to move between the water and the air to reach a state of equilibrium at a predictable ratio depending on the weight of the chemical and the temperature of the water and air. If the concentration of either the water or the air is known, the other may be calculated with a reasonable degree of accuracy if the temperature of the air and the solution are known, and if the ratio of the volumes of the two fluids (water and air) is known, then the total concentration can be calculated from a small sample.

With the Intoxilyzer 5000, the ratio of the concentration of alcohol in the provided breath sample and the concentration of the alcohol in the blood that produced the air value is considered to be 2100:1 and was determined by the assumed volumes of blood and air in lungs, the temperature of the blood, the dissolved alcohol, and the air in the lungs where the gas exchange takes place, and the molecular weight of the ethanol. This ratio is the portion of this process that is, unfortunately, based on fiction, as the ratio and temperatures are not measured in the subject providing the breath sample, but are determined as averages of the general population. In order to enable the instrument to make this assumption and provide a measurement, some value must necessarily be assigned to these variables in order to calculate and provide a blood alcohol reading.

Actual values of the partition ratio as measured in living human beings can run the range anywhere from 900:1 up through 2600:1. There is significant argument among many scientists about the values in this range, but many agree that most subjects will fall within numbers very near to this range of values. Most of the adult population will fall nearer to a 2300:1 ratio. Therefore when using 2100:1 for determining the blood alcohol level, some leeway is being given to anyone that may have an actual ratio in excess of that assumption. Clearly there is some potential for error associated with not knowing the actual partition ratio. Should a test subject have a ratio less than the 2100:1 value, then the test subject's blood alcohol level may be unfairly and inaccurately overestimated above the actual blood alcohol level.

The temperature assumption in the operation of the Intoxilyzer 5000 is the single most sensitive variable involved in making the breath to blood alcohol leap. The instrument uses a test chamber and sample acquisition tube heated to 34 degrees C. The simulator solution is heated to 34 degrees C before being used to provide a "known" concentration of alcohol saturated air to the chamber for calibration purposes. The argument of CMI in designing the instrument and using 34 degrees C is the assumption that the breath leaving the body is at that temperature, and therefore, in their eyes, it makes sense to use this as the temperature on which to base the testing procedure.

It is an extremely important fact that the gas exchange of alcohol across the blood to air interface occurs in deep lung tissue. The manufacturer of the Intoxilyzer 5000, as well as most other evidential breath instruments, has gone to great pains in order to ensure the test sample trapped in the instrument is this deep lung air because of this very fact. Since temperature is the driving force that makes vapor molecules (alcohol) leave the solution and move into the air, it is the temperature at the site of this activity that is most important when attempting to determine how much alcohol will be in the air sample compared to the blood from which it was derived. Gas exchange *does not* occur in the air specimen as the air travels up the respiratory tract to the mouth and into the instrument.

The temperature in the deep lung tissue is going to be the same as core body temperature. This is going to be a minimum of 37 degrees, not the 34 degrees that the test assumptions are based upon. When looking at "normal" body temperature, 37 degrees Centigrade compares to 98.6 degrees Fahrenheit. The warmer the air at the time of the gas exchange, the more alcohol vapor will be transferred to the air from the blood. For many years, it was assumed that 98.6 F was normal temperature, but there have been many research papers produced over the last half century that dispute that fact quite successfully. In reality, there is no "normal" body temperature, as it fluctuates throughout the day, and from day to day. There are a number of factors that influence body temperature, and what may be considered normal.

The most common factor that would influence temperature is time of day. Core body temperature is another biological function that follows a classic circadian rhythm, and as the day progresses, body temperature tends to rise. Temperature is normally at its lowest point early in the morning immediately after rising from sleep. As the length of the wake cycle is extended, core body temperature rises, to a high point late in the evening. Should a "normal" person awaken with a body temperature of 37 C (98.6 F), and then by evening, it would not be unreasonable for the temperature to be 38 C or 38.5 C, or higher. Cellular metabolism is run by energy produced by literally burning sugar, and the more cellular functions that may occur the more heat will be produced. This takes place with every muscle movement throughout the day, from an act as simple as breathing to more energy intense activity such as talking, walking, and any other physical movements. A night time temperature of 100 F to 101 F is no longer considered an indication of fever, and therefore, illness.

Other factors can influence core body temperature to include hormonal cyclic changes in women and men, stress on the body from injury or emotional distress, dietary changes with high protein or high carbohydrate consumption, and other less common conditions such as over or under active thyroid conditions and changes in other endocrine levels (cortisol, insulin, glucagon, etc). The most commonly recognized factor that may change core body temperature is of course illness. Many acute illnesses, even very minor illnesses that may present only as a nuisance, will cause the body, through normal physiological behavior, to produce an elevated core temperature as a disease fighting mechanism. These readings can easily reach levels of 102 F to 104 F.

Experimentation with simulators and the Intoxilyzer 5000 can confirm this information quite easily to show the significant changes that can occur in instrumentation alcohol readings compared to temperature in a standard solution where the concentration is not changed, but the temperature is the only factor altered.

The manufacturer recommends the simulator solution be maintained at 34 C and be mixed to produce a concentration reading of 0.10% at that temperature. A fresh solution should be mixed, and brought to temperature. Base line testing will certainly show a reading of 0.10% if the solution is mixed properly. With some thermostatic mechanism, the solution temperature should then be brought up one degree at a time, and tests should be repeated with the solution at the higher temperatures. A very specific, predictable result will be obtained, and is documented in the chart shown below. These results are reproducible, and will be the same from one instrument to another, with exactly the same readings.

Effects of Temperature on Intoxilyzer Results with 0.10% Simulator Solution

Temperature C	Temperature F	Intoxilyzer Reading
34	93.2	0.100%
35	95.0	0.107%
36	96.8	0.114%
37	98.6	0.121%
38	100.4	0.128%
39	102.2	0.135%

Chart B

As Henry's Law dictates, the temperature of the solution (or blood) will impart more kinetic energy into each individual molecule of ethanol. As temperature of the solution rises, more ethanol will move from the solution into the surrounding airspace. It is not unreasonable to refuse to accept the temperature of 34 C as the proper operating temperature of this system in light of all the information available at this time. The manufacturer should consider a different temperature, or even produce software that will measure actual temperature, and eliminate the assumption of a not only possibly incorrect value, but a very likely incorrect value. All modern research shows that with normal circadian rhythms, hormonal changes, and changes with certain chronic and acute illnesses (diagnosed and undiagnosed) the 34 C temperature should be totally rejected in an attempt to produce correct correlations in breath and blood alcohol levels.

The Intoxilyzer 5000 has certain safeguards built into its operational process and procedures in an attempt to eliminate potential errors in the blood alcohol readings produced. A common interferant present in our bodies is acetone, or ketones, that are a normal byproduct of lipid (fat) metabolism. This chemical is commonly present in any subject that may be burning more fat for metabolic fuel than sugar (diabetics, dieters, any one fasting for any reason). The acetone tends to produce the alcohol-like odor on the breath of diabetic patients, and was even measured as alcohol on some of the older breath testing instruments in the past. In a one month period in the Emergency Department of a large hospital, 32% of all patients tested showed measurable ketones in their urine.

Since the Evidential Breath Test will also measure Acetone as Ethanol, the Intoxilyzer 5000 has a mechanism to measure the alcohol and the acetone in a breath specimen, and then will subtract the value associated with only the acetone from the total, giving the operator a clean, alcohol only reading. Several other alcohols, and alcohol like chemicals, will also cause a false reading on the evidential breath testing equipment, and the acetone is the only chemical interferant that is eliminated from the test procedure. Fortunately, most research indicates if any of these other chemicals are present in measurable levels, they would be fatal to the subject, and therefore, should not be an issue with modern breath alcohol testing.

The simple twenty minute observation period mandated in breath testing on this instrument will eliminate nearly all possibilities of exogenous mouth alcohol that may interfere with the testing process. Most alcohol present in the mouth will be burned off at normal body temperatures before the end of the twenty minute observation period, normally at eleven to fifteen minutes due to the highly volatile nature of the chemical. As an added safeguard, the Intoxilyzer 5000 has programming known as a "slope detector" that will pick up exogenous mouth alcohol. This is done by observing the alcohol reading within the instrument during the breath delivery procedure. Should there be exogenous mouth alcohol present there will be a fast, high initial spike in the alcohol reading that will drop very quickly as all the alcohol is blown through the sample chamber. In this case, the instrument will not give a numerical value for the blood alcohol reading, and will require a new twenty minute waiting period and retesting.

It is understood that there are certain assumptions that ***must*** be used in order to provide a quick, easy, and convenient method to test suspects for blood alcohol levels. In an attempt to produce an accurate instrument designed for simplicity of use, these assumptions are designed into the analytical software of the instrument. The Intoxilyzer 5000 has certain safeguards built in to protect the integrity of the testing process, but caution should be used in that when comparing breath alcohol to blood alcohol, it is not always a clean, simple leap from one to the other. Certain inconsistencies can alter the final result, sometimes to a significant degree. It must be remembered that breath alcohol does not cause any impairment at all, only blood alcohol can be responsible for impairment.

Chapter Four

Gastroesophageal Reflux Disease

An issue with evidential breath testing that is receiving a significant amount of attention now is the problem with GERD, or gastroesophageal reflux disease. This is such a common health concern in the United States that it is estimated that as many as seven to ten percent of the adult population suffers from this difficult condition. It can range in severity from a mild nuisance to a serious, debilitating condition, and occasionally goes undiagnosed for many years in some patients. It is not unusual for a patient to not realize that they suffer from this extremely common health problem.

It is sometimes put out as a defense issue when there is some difficulty explaining a reading on the evidential breath test as unexpectedly elevated for the amount of alcohol actually consumed. It is very common for the prosecution side of a case, either by expert testimony or by argument utilizing several published articles, to simply circumvent this argument and attempt to make it a non-issue. It is usually pointed out that the Intoxilyzer 5000, as well as several other instruments, has a "slope detector" that will pick up the fact that there is mouth alcohol present, and therefore, will protect the innocent from the possibility of an erroneous reading. These "safeguards", as well as the studies designed to prove or disprove this premise, must be closely examined in order to properly draw any conclusion, or more importantly, to not draw an incorrect conclusion.

For anyone to understand how this may be a problem, it is very important to understand the pathophysiology of GERD and the mechanism of the slope detector in detecting exogenous mouth alcohol. There is considerable difference in these two conditions; so much in fact that one has little to do with the other. To argue that someone suffers from GERD does not, and should not, be interpreted as an argument that there was exogenous "mouth alcohol", as these are two different issues, and should never be confused. It is imperative that anyone involved in the defense or prosecution of any case of this nature fully understand the pathophysiology of this disease process. Without a complete understanding of Gastroesophageal reflux disease, exactly how it occurs, and what it can do to possibly affect the breath testing process, it will not be possible to provide an adequate defense (or even prosecution) if one should be warranted.

What is GERD, and Where Does it Come From?

This is a condition that occurs when the junction between the esophagus, which is the tube leading from the mouth to the stomach, and the stomach itself does not close sufficiently to seal against retrograde flow of gastric contents upward. This junction (the Lower Esophageal Sphincter or LES) is a physiological muscular valve with the function of closing the esophagus off from the stomach contents which are essentially being warehoused. Whatever is being contained within the stomach is available for backward flow into the esophagus if the LES begins to malfunction. This can include solids, partially liquidated solids, fluids, as well as gases from anything that may be volatile enough to vaporize. There are many different triggers to cause this backflow, either as an acute episode or in a more chronic situation. Either can be an issue that needs to be evaluated to ensure there is no problem with accuracy of the result provided by sampling of the breath to determine blood alcohol levels.

If the retrograde flow from the stomach into the esophagus is solids or liquids, and is caused by forceful contraction of the muscular stomach walls or other extrinsic pressure inducing stimuli, then this is not "GERD", but is regurgitation. This is one of the important facts to understand when dealing with GERD as a possible issue in the accuracy of the evidential breath test, as the operators are trained to watch for, and even warn the suspect about, regurgitation. Generally speaking, regurgitation is easily detected by both the test operator and the suspect, and can certainly cause exogenous mouth alcohol should there be alcohol present in the regurgitated material. This is particularly true should it reach as high as the posterior oropharynx, or throat.

Many times, regurgitation is argued by the defense (instead of reflux) as a mechanism of causing an abnormally increased Breath Alcohol level, and this is appropriately and easily countered by the prosecution. By arguing regurgitation, they have clouded the issue much more than had they argued the point of GERD properly.

Occasionally, this condition is severe enough that the chronic refluxing of the gastric acid has burned and eroded the esophageal mucosa and causes significant enough discomfort to cause a patient to seek care. Other times the diagnosis is made when a patient presents with complaints of nocturnal heartburn or chest pains, with mild after meal tightness or discomfort, frequent belching or "coughing up food", and even for difficulty swallowing. The symptoms can be so severe as to warrant a very expensive and resource intensive evaluation to rule out a life threatening cardiac condition. One of the most common presenting complaints in a patient previously not diagnosed with GERD is non-cardiac chest pains. There are well documented cases of patients being hospitalized for evaluation of what appears to be myocardial infarction (heart attack) only to find the true diagnosis is GERD. Even though the hospitalization and evaluation may cost upwards of several hundreds of thousands of dollars, symptoms are so dramatic and severe as to make this an appropriate use of resources, and would never be considered to be wasteful or inappropriate.

Some patients present with what they perceive as asthma, allergies, or other respiratory complaints to find after an exhaustive, but negative, respiratory work up, they suffer from GERD. It is a not unusual for a patient to present for multiple visits to their primary care provider for hoarseness, sore throat, ear pain, or chronic sinus pressure and infections, only to be resistant to all attempts at treatment. After finally referring to an Otorhinolaryngologist (ENT specialist), a quick peek into the deep areas of the throat with their specialized scopes reveals irritation and burns classic of acid reflux, and the appropriate referral to a Gastroenterologist can be made to confirm the diagnosis and institute more appropriate therapy.

GERD can affect any part of the anatomy from the lining of the lower esophagus, to the esophageal and tracheal junction, up through the posterior of the throat, and into the ears (via the Eustachian tube) and the sinus cavities. It is not out of the ordinary to see dental enamel erosions and gingival inflammation due to the chronic refluxing of acid and other gastric contents. It is a condition that demands a thorough work up in order to be treated effectively, even if the condition itself is not serious in a life threatening manner. It can mimic nearly any pathophysiological process and needs to be considered by any primary care provider, dentist, allergist or a host of other specialist during any evaluation involving the respiratory system. At best, GERD is a nuisance that may be easily confused with a host of other conditions. It can significantly impact daily functional ability and activities of daily living. At its worst, it can cause serious pain, malnutrition from inability to eat appropriately, to serious mucosal damage with erosion, bleeding, and even death.

The simple fact that this process may present in so many varied ways should be clearly indicative of the possible subtle nature of the disease condition. It should start to become clear at this point how a process with so many presentations may allow for retrograde flow of materiel at almost any time, with minimal ability to predict the frequency, timing, and seriousness of the condition and retrograde flow. This is a condition that can imitate a host of other health problems, and is frequently ignored by many patients until it becomes an issue that begins to interfere in their lifestyles in some manner. It is so varied in presentation that it is easily misdiagnosed by the well meaning practitioner attempting to limit expensive and sometimes uncomfortable testing (frequently appropriately so) by treating the presenting issues before the indications point to the true problem.

It is true that in the early period of this disorder there may be reflux of gastric contents only on lying flat, when wearing tight belts or clothing, when sitting and leaning forward, or after an unusually heavy meal, but after many months of refluxing past the esophagogastric junction, there may be no physical trigger required to allow the more fluid materiel (such as liquid and vapor) to reflux. In fact, there will become a time when there is significant reflux from chemical triggers alone, such as nicotine, caffeine, or, surprisingly enough, alcohol. There are many environmental triggers available on a daily basis to make this a routinely occurring problem for many people. The problem can easily progress to the point where there is no remaining function of any controlling mechanism of the LES to keep anything easily vaporized from entering the esophagus on a regular basis. At this point, the esophagus becomes a reservoir for whatever in the stomach may be easily vaporized.

There are several causative issues that produce the chronic reflux state. In approximately 20% of all GERD patients, their condition is caused by decreased competence of the LES due to decreased LES pressure, increased intra-abdominal pressure due to pregnancy, obesity, or even diagnosed or undiagnosed abdominal masses. Delayed gastric emptying is another, more frequent, cause of chronic GERD. There are many factors that can cause delayed gastric emptying. Gastric outlet obstruction, either from masses or enlarged organs located outside the stomach and small bowel or scarring from previous disease (ulcers, erosions, or tears) will obviously slow emptying of the stomach. Gastric emptying can also be delayed by a host of other issues such as small bowel dysmotility, idiopathic gastric stasis (or simply poor gastric function of unknown source), and a host of neuromuscular disorders diagnosed and undiagnosed. A hiatal hernia is a commonly occurring problem that replaces the lower esophageal sphincter with gastric tissue above the diaphragm, and by its very nature will cause gastric contents to always be within the esophagus.

In all of the conditions that cause GERD by delayed gastric emptying rates, there are two issues that may cause difficulty in reliance on the results of the evidential breath testing instrumentation. First, the refluxant is present for a prolonged period of time, frequently well beyond the normal range of time that one would expect there to be non-absorbed alcohol in the stomach. Secondly, the peak alcohol absorption is delayed and will abnormally affect when a suspect will begin to decline in blood alcohol concentration and when it would be reasonable to expect the result obtained to be reflective of the true BAC at the time of the offense versus the time of the test.

If you are to understand the nature, frequency, and severity of gastroesophageal reflux it is a simple matter to understand how the gastric contents can be in the space between the mouth and the stomach, particularly considering the fact that the pyloric valve (at the far end of the stomach) is usually able to seal more tightly, forcing the contents of the stomach to back up.

Mechanics of GERD and Breath Testing

Alcohol is easily vaporized into a gaseous vapor at body temperature and as we know from our understanding of the partition ratio this is the entire premise behind breath testing to evaluate blood alcohol concentrations. Knowing this fact, any alcohol within the fluid portion of the gastric contents will also be present in the air of the hollow organ cavity of the stomach. Anyone that has ever observed an x-ray study of the abdomen will know that there is almost always a gastric bubble in the upper third of the stomach, even when it is full of food and liquid. It is this gastric air bubble that aids in identification clinically on abdominal x-rays as to position at the time of the films and the location of the organs within the abdominal cavity. This gastric air bubble will reach equilibrium with the alcohol in the stomach contents in the same manner as the alveolar air does with the blood. It is no stretch of any manner to understand that this vapor is going to be present in the esophagus and even anatomically higher in a patient that suffers from reflux.

The breath testing instrument will certainly function on the assumption (appropriately so) that the concentration of alcohol in the breath sample is representative of the concentration in the circulation as calculated utilizing the partition ratio with which it was designed. Since the concentration of the beverage in the stomach is not diluted by the entire amount of total body water, it will be much more concentrated than circulating alcohol being presented to the alveolar surfaces in the lungs. This is true even when diluted with mixers for spirits, and with foods and other consumed liquids. Even when the alcohol consumed is of a relatively low concentration (such as 4% beer), it is still much more concentrated than the alcohol that survives first pass metabolism, and enters the entire water space of the human body. Consider the fact that the evidential breath testing instrument measures an incredibly tiny amount of alcohol, and multiplies this measure by 2,100 to give a concentration of blood alcohol in the hundredths of a percent. Compare this to the amount of alcohol present in a dilute solution of four or five percent (as in beer) that is present in the stomach. Even when this solution is diluted further by what food and liquid could possibly be present in the stomach, the concentration is still hundreds of times stronger than the total alcohol present in the blood, and therefore the airspaces of the deep lung tissue.

When a suspect is given a breath test, he is instructed to blow continuously and forcefully into the sample tube to provide the breath test specimen. As the breath passes the opening at the glottis (where the trachea and esophagus join), there is the potential to pick up vapors from any air or gas within the esophagus. If a suspect has any alcoholic beverage within the stomach at the time of the test, the concentration may be many times the concentration of the alcohol previously absorbed into the circulation, and these hyperconcentrated vapors can potentially be drawn into the air stream of the expired breath enroute to the evidential breath test instrument.

After this lengthy explanation of what GERD is and why it is not unreasonable to expect volatile gastric contents to be within the esophagus, we need to tie into the possibility that these vapors can in fact cause an issue with the evidential breath testing instruments. This is easily explained by an understanding of the physical scientific properties of Bernoulli's principle. Vapors and gases follow scientific laws that have been developed due to the properties that govern their behavior. Unlike the laws of man, these natural laws cannot be manipulated to any significant degree. They cannot be changed simply because man has found them to be inconvenient.

The basic premise of Bernoulli's principle is that in a fluid flow (in our case the fluid is air) an increase in velocity of the flow causes a decrease in the pressure of that flow. As the speed of the moving fluid (liquid or gas) increases, the pressure within the fluid flow decreases. When a fluid is accelerated from a wider space into a narrower tube or constriction, a corresponding volume of fluid must move a greater distance forward in the narrower tube, and therefore will have a greater velocity. The principle of this issue is that as the tube narrows, the speed or velocity increases. As the speed increases, the pressure must decrease, as the velocity and pressure are inversely related according to the ideal gas laws. If one is to move beyond laboratory tubing, reservoirs, and pumps, and picture the lungs, trachea, and the oral cavity, it is easy to see it is the same system.

Progressing beyond Bernoulli's principle, we can be more specific by looking at the Venturi effect, which is a simple modification of the Bernoulli principle. In the Venturi effect there is a feed tube entering into an area of constriction. Therefore, by definition, it enters into an area of lower pressure. If that feed tube originates in an area of normal pressure, and has a source of fluid (again, this would be liquid or gas), then the normal pressure acts on the source of the fluid, and this will promote a flow into the lower pressure and higher velocity of the constricted area that contains the original fluid flow.

There are several common uses of these principles that illustrate this fact quite well. A paint sprayer that operates on compressed air has an air feed from the source that forces air into a restriction in the spray gun. The paint source is in the cup attached to the gun with a draw tube from the cup into the gun's high speed air constriction. Paint is drawn into the air stream and is then sprayed onto the target being painted. Automotive carburetors work on exactly the same principle by drawing air through the throttle body and through a venturi restriction. The fuel is drawn from the fuel storage bowl, is atomized and is then drawn into the air rushing through the throttle body and into the manifold to the cylinders. This air and fuel is mixed in exact proportions due to the restrictions built into the air and fuel metering orifices, and is moved to the cylinders to be burned for energy extraction.

It does not take much imagination to allow one to visualize the human body as a very similar system. The air is drawn into the lungs where gas exchange (carbon dioxide, oxygen, nitrogen, water vapor, and other components such as alcohol) takes place. In the instant just before exhalation occurs, the diaphragm rises, and applies significant pressure to the air within the alveolar sacs within the lungs. The air is forced under pressure from the large area of the lungs into the smaller restriction of the tubular trachea. The air is accelerated and, as we know from Bernoulli's principle, the pressure drops accordingly.

With an understanding of all of these facts presented over the last few pages, it is easy to see that there may be some exposure of gastric contents to the air flow from the deep lung tissue as it passes from the body. As the subject forces air from his lungs, it will pass the glottis and is at that time exposed to a steady concentration of vapors in the esophagus from the gastric fluids. These vapors will be added to any that may already be present in the air being moved to the sample chamber of the evidential breath test instrument. These vapors will also be of a constant concentration and the exposure time will be constant, and will not show the initial spike that drops after the first fraction of a second as would happen with raw exogenous mouth alcohol. Therefore, the slope detector is useless at detecting this additional source of alcohol and warning the operator of a potential and likely error in the provided value.

Mouth Alcohol Detectors

The manufacturer of the Intoxilyzer 5000 has added software to the program that operates the instrument that is designed to detect exogenous mouth alcohol. This portion of the program is known as the "slope detector" and is designed to catch an early high spike of alcohol concentration in the breath sample that falls off early in the sample. It is a simple matter to see that if there is alcohol present in the mouth of a subject, either as a liquid or retained in a foreign body such as food particles, then there may be alcohol vapors presented into the air that passes through the oral cavity and into the sample tube of the instrument. These vapors will show a significantly higher concentration than that of the alveolar air, and will quickly drop to actual breath alcohol levels as the alcohol vaporized from the oral cavity is blown off. This exogenous mouth alcohol source is much different than having a source of pre-vaporized alcohol present, such as that available from the esophagus that is provided over a constant period of time during exhalation.

The manufacturer is correct in their statement that this would actually trigger the slope detector to see an initial spike that drops quickly[1]. It may then drop to regular deep lung air values or may still remain artificially elevated. The company line of the manufacturer is that the slope detector will detect this mouth alcohol and will not give a reading. At that time, the operator should wait another alcohol depravation waiting period and retest the subject, as certainly all mouth alcohol will have evaporated by the end of the second waiting period.

In actuality, the slope detector is accurate and works quite well with some significant limitations. The slope detector is designed to pick up mouth alcohol, and not vapors being refluxed into the breath as it flows from the lungs to the sample tube. After the section explaining the actual physiology of GERD, it is easy to see that anyone who suggests that the slope detector will pick up vapors from refluxant into the respiratory anatomy may not understand the nature of reflux.

GERD cannot be grouped with "mouth alcohol", as there is nothing suggesting that a subject suffering from GERD actually has alcohol passing upwards to the mouth. It can happen, but would be obvious to the subject, and most likely to the operator of the Intoxilyzer performing the pre-test observation period. GERD is *not* mouth alcohol and this cannot be stressed enough. When GERD is the issue, there is no validity to the argument that the slope detector will alert the operator to the presence of mouth alcohol and will abort the test without providing a reading of the actual BAC. There is no mechanism for the Intoxilyzer 5000, in any of its incarnations, to detect if the measurable ethanol in a provided sample is completely from the alveolar spaces, or is contaminated from what would amount to a much higher concentration from the refluxed vapors of the stomach via the esophagus. If protocol is followed, and a second test is run within the next few moments, the two readings obtained would be sufficiently close enough to not invalidate the results of the entire test process.

Current Studies on GERD

When a defense attorney argues the issue of GERD and possible interference with the testing process, the prosecution is usually quick to point out a study performed in Sweden that concludes that GERD is unlikely to alter an actual breath alcohol level to any degree. This study was developed by several well known scientists that have studied breath and blood alcohol testing quite extensively, and was published in 1999 in the *Journal of Forensic Science.* This author has encountered prosecutors and courts that have quoted this study, and used it in court to cast doubt on the defensive argument that GERD may have caused an inaccurate breath test result.

In fact, if one is to read the abstract at the beginning of the article, the last sentence states that the researchers concluded that the "risk of alcohol erupting from the stomach into the mouth owing to gastric reflux and falsely increasing the result of an evidential breath-alcohol test is highly improbable."[2] When first looking at this study and the conclusions drawn, it does in fact appear that this conclusion may be valid. This author has absolutely no argument with that statement except that the issue with GERD is *not* alcohol erupting into the mouth, but is actually the vapors present in the esophagus.

There are several points with this article that cause some concern from a scientific standpoint.

First, this study was developed using a very small study population of only ten subjects. A larger population, although possibly not practical, would have been much more informative and may have provided a much more reliable set of data. Each of these subjects was tested two different times, once with abdominal binding to "induce reflux" and once without, each was tested for blood and breath alcohol levels throughout the experiment, and each was given a measured amount of alcohol on a totally empty stomach after an overnight fast. Prior to the experiment, each of these subjects was found to have incomplete closure of the gastroesophageal junction when evaluated by upper endoscopy.

In the results section of this study, the authors clearly state that breath alcohol levels of some subjects overestimated the venous blood alcohol levels during the first ninety minutes after the end of drinking. During the later times, and throughout the post-absorptive time, the breath readings tended to underestimate the venous blood alcohol levels. The authors explain this difference as expected variations normally seen during the absorptive phase, but they fail to consider the possibility that this difference may be from alcohol vapors present from the GI tract due to this incomplete closure of the gastroesophageal junction seen during the endoscopic evaluation prior to the performance of the study.

What this author finds interesting in this study is that the "usual and expected" difference in breath alcohol levels and venous blood levels occurs in only two of the 10 subjects, and in one of those subjects, the difference was only in the arm of the study with the abdominal pressure applied, and not in the arm without the abdominal binding. The second of the two showed the variation in breath and blood alcohol readings in both arms of the study.

If this variation is truly a simple overestimation of venous blood alcohol levels by breath measurement, then one should expect to see this variation in all, or at least most, of the subjects. Also, one should expect to see this variation in both arms of the study, and not only in the arm that involved the binding of the abdomen to induce reflux. The fact that the elevated breath readings are present only during the absorption phase might possibly be due to the fact that there is alcohol present in the stomach that has not yet been absorbed. With the alcohol present in the stomach, there certainly will be vapors present in the esophagus of a person with incomplete closure of the gastroesophageal junction. It may be possible that the conclusion of these researchers is flawed, and with the facts as presented (not present in all subjects, and present only during the time the alcohol is in the stomach) there may be further study needed to ensure the provided conclusion is accurate.

These researchers state that the "differences between BAC and BrAC observed during different stages of the pharmacokinetics of ethanol did not seem to depend on whether or not reflux was provoked"[3]. This is clearly in agreement with the facts presented in this text that reflux is not an issue that comes and goes, but in fact can, and usually is, present all the time. It is certainly not dependant on any particular act or mechanical device to "provoke" reflux. When a subject suffers from incomplete closure of the gastroesophageal junction, it is a chronic condition that may be influenced by acute triggers, but it is always present. A subject suffering from GERD suffers from GERD. It is not a condition that may or may not be present during the evidential breath test, but is a condition that is always present until treated and/or cured.

A second paper was prepared by Dr. Alan Wayne Jones (one of the authors of the previous study) on 16 August 2006. In this second paper, Dr. Jones references a conference at which he spoke in May 2006 for the California Association of DUI Defense Lawyers. During this conference, Dr. Jones pointed out that the evidential breath testing instrumentation currently in use might not be able to detect mouth alcohol under some circumstances. Each of these instruments is fitted with a slope detector in the program that is designed to detect mouth alcohol, and prevent the analysis from providing an erroneous reading. These slope detectors are designed to monitor the rate of change in the alcohol concentration of the expired air as it enters the test chamber for analysis. The assumption is that if the alcohol concentration spikes in the initial aspect of the expired breath, and then drops suddenly, then there could be exogenous mouth alcohol present.

Dr. Jones states that "there is no published evidence that the more dangerous form of mouth alcohol, namely that which might erupt from the stomach in connection with a burp, belch or regurgitation can be successfully detected and distinguished from alcohol originating from the lungs."[4] It has been noted that many times the slope detectors can malfunction, and completely fail in their intended purpose.

The slope detectors are checked by the manufacturer, and even during classroom instruction on instrument operators courses, by having a subject swish a solution of whiskey (or other spirits) in the mouth, and then having that subject provide a breath sample. In nearly every instance the instrument will certainly detect the exogenous alcohol, and will notify the operator of this fact, preventing an inaccurate result.

If the test subject were to continue providing samples over the next 15 minutes, a predictable pattern of results would appear. For the first several samples, usually for the first 4 to 5 minutes, the result may be invalidated by the slope detector as having mouth alcohol. Surprisingly, at a period of about 7 to 10 minutes after the exposure to this strong solution, the slope detector will usually fail, and will provide a result, giving the indication that the alcohol measured in the breath sample, corresponds to a blood alcohol concentration, even when there is essentially ***no*** alcohol in the blood at that time. As expected, after the standard 15 minutes, the residual mouth alcohol would have been totally eliminated, giving an expected result of no alcohol detected.

Studies performed by several different investigators, to include Dr. Jones, this author, and several other noted alcohol researchers, prove this pattern consistently. This is particularly true when the subject has some alcohol in his system. The low concentration of blood alcohol seems to make the failure of the slope detector more likely than even in a subject with a BAC of zero percent. In his article of August 2006, Dr. Jones describes a test run on a subject who had an actual blood alcohol concentration, then swished with an alcohol containing mouthwash. Subsequent tests at 8.5 minutes, 10 minutes, and 17.5 minutes showed three very different BAC readings, and none were invalidated by the slope detector in the instrument. Clearly, the reading at 17.5 minutes was a true blood alcohol concentration level, and the two different (and successively higher) readings taken closer to the exposure to the mouthwash ***must*** have been erroneous readings.

This author has observed this exact same phenomenon many times. One DOT approved instructor on the Intoxilyzer 5000 showed time and again that the slope detector repeatedly failed to detect exogenous mouth alcohol, in both subjects with a low blood alcohol concentration, and less frequently (but still noted) in subjects with a zero percent blood alcohol concentration. These results were reproduced using several different subjects over the period of two days of testing, to include the instructor himself as well as other test subjects of both sexes, and with and without diagnosed GERD.

Considering the fact that the alcohol solution within the stomach of a subject arrested for DUI will be less concentrated than the alcohol contained in the straight mouthwash or whiskey, then there can be no doubt that the slope detector on the current generation of evidential breath test instruments may potentially miss the fact that some of the measured alcohol may be present from gastric vapors, and not the alveolar air as assumed. This is an error that may easily mislead the trier of fact into a conviction based on an inaccurate reading on the breath alcohol certificate.

GERD Conclusion

Alcohol from refluxed vapors is only a possible issue if there has been a relatively short period of time elapsed from the time of the last drink to the test to allow for alcohol to be present in the stomach. Once all of the alcohol has passed to the small bowel, there clearly can be no alcohol present to reflux into the breath sample passing across the esophageal opening and into the instrument. Any argument that a subject may have an inaccurate test result due to reflux when that test was given beyond a reasonable time for complete absorption is an argument with no merit. For reflux to be considered an issue, the alcohol must still be present in the upper GI tract (meaning the stomach or esophagus).

With all of this information available, it is clear that more in depth studies need to be performed in order to determine if the problem of Gastroesophageal reflux disease is an issue that may cause problems with the validity of the evidential breath testing equipment now in use. Manufacturer's claims that the slope detectors are capable of eliminating alcohol from a test unless that alcohol is a part of the breath sample from the lungs is clearly lacking in any form of scientific evidence or proof. The fact that the government and prosecutorial experts simply parrot the statements of the manufacturers needs to be addressed with newer and better designed studies. These studies will need to have larger numbers of subjects, and will need to have several arms in order to show once and for all if the slope detectors are capable of delivering as promised by the manufacturers of these instruments. Only then can we be certain that any measured alcohol gives a true reading of the defendant's circulating blood alcohol, and does not include any alcohol from any other source.

Most importantly, it is important to remember that reflux is not exogenous mouth alcohol. These are two entirely different issues. Also, for alcohol vapors to be available to the exhaled breath to be carried to the instrument, there does not have to be an actual diagnosis of GERD, nor does the patient have to have physical evidence of GERD at the time of the test in order for there to be a potential for a falsely elevated blood alcohol reading. It is important to maintain the integrity of the testing procedures and the confidence of the results that all possible errors be eliminated before bestowing life altering criminal convictions on defendants based on the results of these tests.

Chapter Five

Standardized Field Sobriety Testing

Introduction

When attempting to determine levels of impairment in operators of motor vehicles, law enforcement officers today need every possible tool available in order to develop the case in order to ensure convictions when appropriate. The enforcement of DUI laws in the United States today is an extremely scientific based process, and requires considerable knowledge of several areas of physiology on behalf of law enforcement officers, prosecutors, and defense attorneys in order to ensure appropriate court outcomes at trial.

The application of scientific principles is no more important in any area of DUI enforcement than in use of the standardized field sobriety tests. In order to allow for a reliable battery of tests to evaluate impairment, the federal government has undergone extensive expense and time expenditure in this endeavor. The standardized field sobriety tests (hereinafter called "SFST") were developed to allow for officers to have a reliable battery of tests that have undergone significant scientific scrutiny, peer review, and repeated evaluations for effectiveness.

In order for any scientific test to be considered forensically reliable, it must be shown that the tests being utilized have been shown to have a reasonable reliability and can give reproducible results when used in different conditions. Polygraph testing is one area that has undergone reliability evaluations, and even though these tests may give 85% to 90% accuracy, they usually are not permitted in criminal proceedings as there may be some level of misjudgment when looking at these results.

To provide a useful tool for DUI enforcement efforts, it was determined that a group of tests would be needed in order to give credence to law enforcement officers attempting to provide valid probable cause for arrest and determination of impairment. In cases of refusal, these tests should be able to provide some level of indication of impairment in order to allow the court to determine whether a driver may have been impaired, even in light of the absence of other scientific blood alcohol measurement results.

Traditionally, there have been geographic differences in types of roadside testing that were being used. Some of these tests were self designed, others were taught in police training programs, and some were just handed down from one generation of officers to the next. None of these early tests were evaluated for scientific reliability until the US Department of Transportation stepped in and the National Highway Traffic Safety Administration (NHTSA) decided to "standardize" these tests, and provide a level of predictable reliability to enable use in courts for indicating impairment.

DUI cases more often hinge on scientific evidence than most other commonly tried cases. This is an offense steeply based in science, from the stop, to the determination of probable cause for arrest, and for determination of actual blood alcohol levels for presentation in court as evidence. A certain level of scientific knowledge and understanding is necessary in every step of this process. Convictions must be based on proof of the basic element of the offense—alcohol impairment. This determination is directly based on scientific research and evidence.

It is mandatory that law enforcement officers, attorneys for both the prosecution and defense, and the courts understand the science behind these tests in order to identify any flaws that may lead to inappropriate convictions or acquittals for what has become a serious criminal offense. The lack of compliance in application of the testing process allows for the possibility of inaccurate conclusions and the possible conviction of an innocent person for what has become a crime with significant financial and social ramifications.

Making a proper DUI arrest has several very important steps. Once an officer determines a driver needs to stopped and evaluated, the most difficult part of the case has just begun. These tests must be used exactly as prescribed, and administered in as ideal of conditions as possible, and with a complete understanding of the findings on observation. The guidelines determined by NHTSA are exact, and must be followed exactly as given in the training manual in order to allow any viable conclusion. Any deviation from these guidelines in any form can totally invalidate any conclusion drawn, and may jeopardize an otherwise excellent case, or may convict someone unnecessarily.

History of Standardized Field Sobriety Tests

The National Highway Safety Administration (hereinafter know as NHTSA), an arm of the United States Department of Transportation, sponsored research in the early 1970's to develop scientific validity to a battery of tests that would allow law enforcement officers a method to determine impairment in drivers suspected of intoxication. The purpose of this research and the development of these tests were to standardize a testing method that an officer could use that would allow for a reliable conclusion in all cases. The NHTSA commissioned a group of scientist under their guidance and direction to evaluate the various "field sobriety tests" being used in DUI enforcement at that time The NHTSA sponsored research was eventually successful in determining twenty standardized "clues" to assist law enforcement officers in the field to identify drivers with blood alcohol levels at or above 0.10% in order to maximize an officer's time in stopping only those drivers most likely to be impaired, and therefore, a hazard to public safety. Since this original research, additional research has assisted by developing clues for determination of impairment in the lower range of 0.08% and for operators of motorcycles.

Prior to the development of these tests, officers generally used regionally and locally developed tests either self taught or passed on from senior officers during training programs. Some of these tests being utilized have since been proven to have absolutely no value in predicting any level of alcohol impairment. Some of the tests being used actually could confuse the matter to the point of causing the false determination of impairment and, therefore, an arrest based on nothing more than minor suspicion due to the "odor of alcohol" and some driving infraction. Even during this period of time, some officers and departments had stumbled onto tests later shown to have reasonable reliability in predicting impairment and justifying an arrest and eventually a conviction, even in light of no chemical test for impairing substances. It was this set of tests that needed to be "standardized" and disseminated to other officers and departments across the country.

The US Department of Transportation, and NHTSA, had several questions that needed to be answered. They needed to determine if a certain test was effective in determining impairment, whether it was nothing more than an excuse for an arrest and blood or breath test, or if it was nothing more than a measurement of nervousness on behalf of a driver stopped by the police after only a drink or two. It was this set of questions that drove the government to develop this set of "Standardized Field Sobriety Tests" discussed and used so commonly today.

Early Research--Laboratory Setting and Field Application

In 1975, the US DOT, through the Southern California Research Institute, approached the Los Angeles Police Department, and was given the use of their Traffic Division for the early lab phase of the research into standardized field sobriety testing. Dr. Marceline Burns and Dr. Herbert Moskowitz conducted the first phase of the research into the road side testing for impairment. During this same time period, studies were published in Finland that referenced the Horizontal Gaze Nystagmus (HGN) test as a predictor of blood alcohol levels. Previously, in 1974, researchers Wilkinson, Kime, and Purnell published a report showing consistent HGN changes with increasing doses of alcohol.

During the first phase of study in Los Angeles, six different roadside tests were evaluated. In addition to the HGN test, the one leg stand test, finger to nose test, the walk and turn test, and a pencil to paper tracing test were studied. Three of these tests were found to have a high level of reliability as a tool for assisting law enforcement officers to make a decision to as to whether a suspect may have been at or above the blood alcohol level of 0.10%. The Horizontal Gaze Nystagmus Test, the Nine Step Walk and Turn Test, and the One Leg Stand Test were found to constitute a potentially reliable battery of tests for determining a BAC of greater than or equal to 0.10%. Observation of certain behaviors during the performance of these tests became "clues" that the officer could utilize to evaluate the driver during a roadside investigation into impaired driving.

Drs. Burns and Moskowitz published their report in June, 1977 as *Psychophysical Tests for DWI* (NHTSA Report No. DOT HS-802-424). This report provided documentation of their findings from the initial phase of research, and their recommendations for further studies evaluating the field use of these particular tests for determining the impairment of suspects from DUI traffic stops.

This second phase of research was initiated almost immediately, and was designed to validate and standardize these three particular tests and develop them into a useful tool for law enforcement officers across the nation. The next report, *Development of Field Tests for DWI Arrests* (NHTSA Report No. DOT HS 805-864), was published in March, 1981 and was authored by Drs. Burns, Moskowitz, and Tharp. This new report provided the details of the reliability studies involving the now standardized tests when used individually or in conjunction with the each other for determining potential impairment.

It is important to remember that at the time of these research programs and the development of the protocols, most states had the level of 0.10% concentration of blood alcohol for determination of impairment. These levels of reliability were documented at that level, and when the tests were given under ideal situations and exactly as prescribed. Reliability levels for each test are listed in Chart C.

Standardized Field Sobriety Test Reliability Chart

Field Sobriety Test	**Reliability Level**
Horizontal Gaze Nystagmus	77%
9 Step Walk and Turn	68%
One Leg Stand	65%
Three Combined Tests	80%

Chart C

An extremely important, and frequently overlooked, point in this early developmental stage is that all of this experimental data was collected in ideal, laboratory conditions. The above reliability levels are often quoted by both the defense and prosecution in courtroom trials, and rarely is that point brought out by either side. Also, the researchers made certain to make the point that the tests must be given exactly as instructed in order for there to be any confidence in the conclusion drawn from the performance on these tests. Any variation, no matter how small, would provide the potential to completely invalidate the conclusion of impairment or sobriety.

The third phase of research sponsored by NHTSA took place in 1983. This research protocol was developed to evaluate the standardized field sobriety tests under field conditions by actual law enforcement officers. The three test battery was field tested by four large municipal police agencies in the eastern United States. There were three objectives of this third phase of research and evaluation.

First, any test used needed to be practical for each and every traffic stop in which the motor vehicle operator was suspected of being under the influence of alcohol. It needed to be effective in selecting the proper steps in managing the next step the officer needed to take in making an arrest or no arrest decision. The first objective was to develop standardized, practical, and effective procedures in reaching this decision.

The second objective in this study was to test the feasibility of these new procedures under the operational conditions of the real-life law enforcement world on the street. These tests must be designed to ensure the officer could use the tests without putting himself in any undue danger. He would be able to make the proper decision and get back to work in short order, either processing a DUI arrest or getting back to patrol duties, as quickly as possible and as safely as possible.

It was absolutely mandatory that the results obtained in the laboratory be reproducible in the field, so that the law enforcement and legal communities would have the confidence to use and interpret the results of these standardized tests. Because of this requirement, the third objective was to secure enough data to compare the field results with the earlier laboratory results, and give the confidence to make the leap from lab to the real world of in the street law enforcement.

After completion of the third phase, the NHTSA staff collected all of the field data from the officers of the chosen police departments and reviewed and catalogued the data with a view toward systematizing the administration of the three tests. It was their intent to ensure that the three test battery was quick and easy to administer, regardless of the career and experience levels of the officers in the field. The tests must be easy to use to identify operators with a BAC at or above 0.10% and minimize the contact with operators whose BAC was below that level.

These standardized elements studied became the three test battery of standardized field tests with specific clues to be observed and reported. It was this set of clues that were used to determine which drivers may be impaired, and they became the criteria by which the arrest/no arrest decision were made. When these standardized elements were utilized, the tests were found to be "highly reliable" at BAC's in excess of 0.10%. The final report was authored by researchers Anderson, Schweitz, and Snyder. It was titled *Field Evaluation of a Behavioral Test Battery for DWI* (NHTSA Report No. DOT HS 806-475) and was published in 1983. It summarized the field application studies of the standardized elements of the tests.

Finalization and the Birth of SFST

All of the NHTSA reports were presented to the Advisory Council on Highway Safety of the International Association of Chiefs of Police (IACP) in 1986. This council passed a resolution that all law enforcement agencies should adopt and implement the new Standardized Field Sobriety Tests (SFST) as developed by the National Highway Traffic Safety Administration. In 1992, the IACP Highway Safety Committee recommended development of training standards for this system of nationally accepted testing procedures for evaluating drivers suspected of DUI. This committee developed and published the standards used today for training law enforcement officers in the use of SFSTs.

Due to the ever changing world of law enforcement, research into the use of SFST continues today. Virginia, as well as many other states, has lowered the legal limits of impairment from 0.10% to 0.08%. Since all the original research was directed at the impairment level of 0.10%, these same tests needed to be standardized to the new limit. Most of the original research was never peer reviewed, and therefore was not subjected to the argument and scrutiny of the scientific community. All of the original scientists have even continued to perform ongoing research into these tests, and have published consequent articles that indicate there may be some limitations to the use of these original SFST and there may be some flaws in the conclusions of the original research. The original research published by NHTSA indicates a level of reliability for a limit of 0.10% that isn't near 100%. When attempting to use this same research and set of tests for other BAC's less than the originally documented level, the SFST simply can not be relied upon for determination of the proper conclusion.

In light of this question, new research in the last several years has attempted to validate the SFST for making arrest/no arrest decisions at the lower 0.08% blood alcohol level. This research documents a study based on information obtained from the drunk driving enforcement squad of the San Diego, California Police Department. On the surface, this research appears to validate the SFSTs at 0.08% versus the voluminous research into the 0.10% level (research that at the time clearly indicated the limits to the usefulness of these tests below the 0.10% level).

It could well be that the reasons these officers performed so well in making the proper decisions below the 0.10% limits was that they are so highly trained and experienced in detecting impaired drivers. This squad of specialized officers has some of the most highly trained law enforcement officers in the United States at detecting and arresting impaired drivers. The correlation to determine blood alcohol levels below 0.10% may not have been quite so obvious had the study used less specialized officers. It is natural that in order to evaluate a specific process, the highest trained, most specialized officers, with the greatest experience should be used. This fact should be remembered when looking at the data, and understand that this battery of tests is used every day by police officers of every level of training and experience, from the brand new rookie cop on the street to the old timer ready to retire. This very interesting study is dealt with in much more detail later in this text, and the specific flaws are highlighted to allow a clear view of the difficulty in resetting the impairment level for proper conclusions.

The **Standardized Field Sobriety Tests**

After this lengthy introduction to the development of these field impairment evaluation tests, it is necessary to finally explain the prescribed tests, exactly what they are, and how they are to be utilized in order to allow the law enforcement community to make the necessary and appropriate determination for arresting impaired motor vehicle operators. The name given is itself descriptive of the tests themselves. They are ***standardized***. This means that there is a method to their performance and use. It is the same in California or Virginia, Maine or Florida. A law enforcement officer can go from one geographic area to another, and the same tests will be used and the information obtained will be the same. The same officer in the same jurisdiction will be able to reproduce the same tests from one day to the next, one month to the next, and one year to the next, all with the same ability to draw a conclusion from observing the performance on the tests. This is due to the fact that they are standardized, and will have reproducibility. The training emphasizes the standardization of these tests, and the manner in which they are used. Looking at the name further tells us they are ***field*** tests. They are not used in a laboratory or office setting, but in the imperfect location of the real world, or in the "field".

The Standardized Field Sobriety Tests (SFST) used in the enforcement of the DUI laws are the Horizontal Gaze Nystagmus test, the One Leg Stand test, and the Nine Step Walk and Turn test. Each of these tests is actually a modification of medical tests used by health care providers on a daily basis in order to evaluate patients and diagnose health issues and illnesses. These tests, whether used by medical practitioners or law enforcement officers, require extensive training in physiology, both normal and abnormal, in order to draw the proper conclusion from the results obtained.

In a clinical setting, these tests are given in a very controlled environment, and this is extremely important for using the results of these studies for developing any meaningful diagnosis. These tests are given in a warm, well lighted, quiet exam room to allow the patients to perform to the best of their ability, and ensure that any missteps are due to physiological issues and not environmental. There should be no outside distractions of any nature, whether it is sounds, temperature or lighting discomfort, or physical irregularities of the test room. The patient must have the utmost ability to understand the directions of the testing procedures to ensure fully capable compliance to the best of their physical ability. Medical training guidelines demand strict adherence to testing environment in order to draw any reasonable and valid conclusion.

Part of the training of medical practitioners is to educate the provider to understand the many, many possible reasons for failures in the performance of diagnostic testing. A test result may potentially indicate a neurological dysfunction of a peripheral nature, a neurological dysfunction of a central nature, or something as wide ranging as drug, alcohol, or other toxic exposure. The potential for metabolic abnormalities that could cause variations in test responses is something that must be understood and considered when observing these diagnostic results. A patient may have a genetic abnormality that could cause a problem with function of the central or peripheral nervous system, even if that disorder would not manifest itself until much later in life. Chronic exposure to environmental toxins or chemicals may cause nerve or other organ damage, permanent or temporary, that may limit function or ability on certain neurological tests. Strokes, tumors, or trauma (even old past trauma) may cause the patient to not be able to pass certain tests for neurological function.

These abnormal behaviors must be evaluated whether the testing is being performed in a medical setting for diagnostic purposes, or in the field for sobriety determination purposes, and although the situations are somewhat different, each will have dire consequences if misinterpreted. Since the law enforcement officers are using modifications of these medical diagnostic tests, failure to properly instruct the client in expectations, and evaluate the results of the performance, will nearly always invalidate the results and, therefore, the conclusion drawn. The warm, comfortable, quiet, distraction-free medical office exam room is a far removed situation from the busy, bustling, noisy, hectic street side observation area in which the field sobriety tests is normally given. This is quite unfortunate, but at the same time, is absolutely unavoidable.

Horizontal Gaze Nystagmus

Nystagmus is a rhythmically repeating oscillation of one or both eyes in different fields of gaze. It is involuntary, and may be present in one or more fields of view. These movements may be classified as smooth and flowing, as a pendulum would swing, showing equal speed and amplitude in each direction, or they may be jerky, more in one plane, less in another and different rates and amplitudes from one direction or side to the other. There could be a slow movement to one side, followed by a quick jerky return to the neutral position. Medically, nystagmus is understood to be a process involving the central nervous system, the eyes, and the organs of balance (the semicircular canals of the inner ear). These organs of balance, and the neurological connections, involuntarily control the movements of the eyes in an attempt to keep them level and in alignment with the head and the central nervous system. It is this function that causes the appearance, grade, and type of nystagmus depending upon different issues.

Nystagmus is listed as Grade I, Grade II, or Grade III, depending on when and where the nystagmus appears. Nystagmus is Grade I when the oscillations appear when the eyes are deviated toward the rapid component; it is Grade II when oscillations are present with the eyes in the neutral position; it is Grade III when the oscillations appear when deviated toward the slow component. These oscillations may be horizontal, vertical, oblique, rotational or circular, or even any combination of two or more of these. The directions of the oscillations will change with reference to the direction of the gaze while observing for nystagmus.

When describing nystagmus, amplitude of the oscillations refers to the degree of the movement, and the rate refers to the frequency or speed of the oscillation. In almost all cases, the faster the rate, the lower the amplitude of the individual oscillation.

Physiologic Nystagmus

Physiologic nystagmus is one of the three main types of nystagmus. It can be elicited in a normal, healthy subject, and will have no pathology or reason for its appearance other than being a normal variant.

End point nystagmus is an oscillation that occurs only at the extreme lateral deviation of the eyes, and will generally appear at approximately thirty seconds after complete deviation. It will appear sooner in a normal person and will be more pronounced when the subject is overly stressed or fatigued, such as would be found in a DUI suspect stopped late at night.

Optokinetic nystagmus is found in normal subjects when the eyes are exposed to fast moving objects (thus the "optokinetic") in the line of sight. This can be related easily to a DUI suspect on the side of the highway with fast moving vehicles passing through the line of sight. A suspect performing the HGN test will be attempting to focus on the light in the officer's hand while his eyes may be involuntarily tracking the quick passing headlights or taillights of traffic.

Caloric stimulation of the semicircular canals is the third type of physiological nystagmus. This is normally not going to be an issue in law enforcement except when a subject is removed from a warm vehicle, and there is a cold precipitation (rain or snow) falling. Should a cold drop get into the ear canal and contact the tympanic membrane, there will be nystagmus noted.

None of these forms of nystagmus are in any way related to alcohol or drug impairment, yet each can be observed during a traffic investigative stop for DUI. It would take significant training and experience to enable an officer to determine the fact that the nystagmus is due to a blood alcohol level, and not simply a physiological nystagmus. As explained, it is not out of the realm of possibility for any of these three types of physiological nystagmus to be present. As the name implies, this type of nystagmus is a normal, physiological process in an otherwise healthy, sober individual.

Pathological Nystagmus

Pathology is a medical term that indicates some form of a disease process, or abnormal functioning in a living system. Pathological nystagmus is present due to some disease state or abnormal physiological process that elicits the oscillations. There are six different categories of pathological nystagmus: Congenital, Neurological Disturbance, Vestibular, Gaze, Hysterical, and Spasmus Nutans. Some of these are extremely rare, and are unlikely to be encountered by law enforcement officers in the field, but some are quite common and can be easily confused with alcohol induced HGN as part of the field sobriety evaluation. It is not enough to ask a subject on the roadside if they have any medical or physical condition that may cause nystagmus, as not all pathological conditions may be known to a DUI suspect at the time of the traffic stop. There is the potential for as yet undiagnosed health issues and disease processes that could be causing nystagmus that could permit an improper, yet unavoidable, error in interpretation of the HGN test. Nystagmus secondary to a neurological disorder can be an issue at any age, and therefore, may be an interfering issue when evaluating a DUI suspect.

Vestibular nystagmus is one form of jerky nystagmus that is not too uncommon and may be encountered frequently by law enforcement officers while observing suspects performing the HGN test. This is associated with some form of pathology and dysfunction in the vestibular system, or inner ear. The physiology of the inner ear is two fold. The inner ear has the primary purpose of transmitting sound waves converted to electrical energy to the brain for sound interpretation. It also has the secondary function of balance, and has the potential through this function of causing a readily visible nystagmus.

In most cases, vestibular nystagmus will also be associated with other medical difficulties such as tinnitus (ringing or hissing in the ears), vertigo (imbalance and nausea similar to "seasickness"), and even decreased hearing sensitivity. Several somewhat common viral infections (the common cold for one) can contribute to the nystagmus associated with the vestibular system. Other causes of vestibular dysfunction are encephalitis (infection of the central nervous system), multiple sclerosis, trauma to the inner or middle ear, or neoplastic processes (tumors, either benign or malignant). Meniere's disease is an illness commonly seen in the medical community that presents as a generalized dizziness and chronic nausea, normally positional in nature. Its cause is very often never located, and treatment consists of only symptomatic care. Vascular lesions of the inner and middle ear could be related to stroke, arterial blockages, or other inflammatory issues, to include migraine headaches. Labyrinthitis is a generic term for any inflammatory process of the balance control center of the inner ear.

While some of these causes should be obvious to the suspect and the law enforcement officer evaluating the suspect, some are quite difficult to detect, even for the suspect and their health care provider. It is not uncommon for a patient to present to their doctor on multiple visits before the true diagnosis is determined. Many of these illnesses are considered a diagnosis of exclusion, meaning there is no specific test to confirm the diagnosis, but the clues are put together and more definitive causes are eliminated from the differential diagnostic list.

Positional alcohol nystagmus has two different components, both of which are related to altering the specific gravity (a measurement of concentration) of the blood by dilution with alcohol consumed and absorbed. The vestibular fluid within the inner ear is equal in concentration to the fluid portion of the blood, as that is where the components of the vestibular fluids are drawn. When the specific gravity of the blood changes abruptly, the vestibular fluid lags behind for a short period of time. As the arterial blood alcohol concentration is rising, the blood will have a lower specific gravity than that of the vestibular fluid. Similarly, as the blood alcohol level begins to decline, the vestibular fluid will have a lower specific gravity than the blood. It is this inequality of the concentrations that will cause nystagmus and vertigo-like balance problems. The difficulty with this form of nystagmus is that it is ***not*** the nystagmus observed with Horizontal Gaze Nystagmus of SFST fame, but is nearly impossible to differentiate.

Gaze nystagmus is another form of pathological nystagmus, and will be the most common form encountered by law enforcement officers. This is the nystagmus of DUI enforcement, and can have many causes, only one of which is alcohol impairment. Causes vary, and include such a wide array of possibilities that care must be taken to ensure what appears to be nystagmus from alcohol impairment is not directly related to other sources.

Nystagmus is such a complex issue and phenomenon that even highly trained health care professionals and medical researchers have some difficulty with categorizing the different forms. Without complex, and frequently computerized, equipment, it is nearly impossible to identify the form and source of observed nystagmus. Any time any nystagmus is detected without obvious cause, an involved and lengthy diagnostic medical workup is warranted to identify the source and potential pathology behind the nystagmus.

In summary, nystagmus is not a definitive indication of alcohol use and impairment, as the Standardized Field Sobriety Test training manuals would lead a law enforcement officer and court to believe. Nystagmus may be present in a normal, healthy, sober individual and could be related to something as benign as fatigue, headache, or even a mild common cold. It could be present as a component of some as yet undiagnosed medical condition, some common prescription or over the counter medication, caffeine, or even exposure to nicotine. Alcohol is simply a very small player in the observation and evaluation of nystagmus.

Nystagmus in DUI Detection

The Standardized Field Sobriety Test training manual describes three general types of nystagmus. This is an over simplification of the explanation of nystagmus and these three general forms of nystagmus of the SFST manual are covered by the in depth explanation provided within this text. Although the training manual breaks nystagmus into more simplified categories to make it easy to understand for DUI training purposes, it closely mirrors the earlier, more complex definitions. With this earlier very convoluted explanation of the different and varied forms of Nystagmus, it certainly isn't expected that one would understand all the fine points of this complicated issue. Hopefully, it will at least open the reader's eyes to the incredibly complex nature of this very oversimplified Standardized Field Sobriety Test.

Horizontal Gaze Nystagmus (HGN) is an involuntary jerking of the eyes as they approach maximum deviation to the sides of the range of the gaze. This being 'involuntary', the person being tested is totally unaware that the eyes are exhibiting this oscillation and they have no control over this movement. As this is a completely naturally occurring phenomenon, the subject can neither enhance, nor prevent this nystagmus. It is extremely important to note that alcohol and/or drugs ***will not cause nystagmus,*** but only ***enhances*** nystagmus that may already be present, even though officers are frequently heard to testify in court that the nystagmus was caused by the presence of alcohol. ***This fact is very specifically noted in the training manual developed by NHTSA,*** but is ignored more often than not.

Other than the medical causes of nystagmus previously detailed, nystagmus can also be caused by extreme fear or anxiety. This is probably going to be present in any case where a subject is stopped by the police, particularly after consumption of even small amounts of alcohol. This anxiety may be compounded should the encounter be precipitated by a motor vehicle accident. Environmental allergies are commonly treated with antihistamines, both prescription and over the counter, and these medications as a class are central nervous system depressants. These can also cause nystagmus changes by exactly the same mechanism of alcohol. Asthma is one of the more commonly treated serious health issues, and the most common medications to treat this (albuterol, theophylline, and singulair) will also enhance nystagmus. Hypertension, or high blood pressure, is one of the top medical diagnoses encountered in adult medicine today. This condition, as well as several different classes of medications to treat it, will certainly enhance nystagmus by several potential pathways. High doses of nicotine and caffeine absolutely not only enhance nystagmus, but does it to such a significant degree that it is noticeable even without close observation.

Ironically, most traffic stops occur late at night (fatigue and caffeine), cause significant duress on behalf of the suspected impaired driver (stress and anxiety), and probably involved activities at a smoke filled bar or restaurant, even if the suspect wasn't actually smoking. This sets the stage for nystagmus, even in a driver that may have had minimal alcohol consumption, if any at all.

NHTSA researchers have documented in their report "Psychophysical Tests for DUI Arrests" that the "...onset of nystagmus at about 40 degrees would correlate with a blood alcohol level of 0.10%". In order for this to be true, the Department of Transportation, through NHTSA, requires that the tests be performed *exactly* as instructed. All of these field sobriety tests are valid only when "properly done" and any deviation from the exact procedure will render the tests ***totally*** useless.

To properly perform the HGN test, the officer is required to hold some focus stimulus (usually a pen or light) twelve to fifteen inches from the face, and move it in an arc (in order to maintain the same distance from the eyes through the movement) across the face of the suspect at a rate of twenty degrees per second. There are six different clues to determining impairment, three on each eye. The officer administering the test looks for smooth tracking of each eye while following the stimulus, the officer observes the angle at which the nystagmus first appears as the eyes are deviated laterally (looking for onset before 45 degrees), and the presence of nystagmus at maximum deviation are the complete set of clues, with four indicating a potential blood alcohol level of 0.10% or more. When performed properly, this test should yield a 77% chance at detecting a suspect with a blood alcohol level of 0.10% or greater.

Horizontal Gaze Nystagmus Clues

Lack of Smooth Pursuit
Onset of Nystagmus Prior to 45 Degrees of Deviation
Nystagmus at Maximum Deviation
THE PRESENCE OF FOUR OR MORE CLUES INDICATES A POSSIBLE BAC OF 0.10% OR MORE
The above clues are to be counted for each eye, and therefore there is a maximum of six clues

Chart D

Even though the instruction manual of NHTSA clearly states that the angle of onset of nystagmus of 45 degrees corresponds to a BAC of 0.10%, the actual research performed under the direction of NHTSA indicated that the angle in question "averaged 41 degrees and decreased to 36 degrees after midnight". In 1981, Tharp, Moskowitz, and Burns (the authors of the NHTSA report) cautioned that there may be as high as a 0.04% error in determining BAC levels. Even with this documentation, NHTSA has declined to accept any recommendation for change, or even mention this variation, in any of their training manuals. Very few law enforcement officers, attorneys, or Judges therefore even know that the DOT sponsored training manual purposely ignores potentially exculpatory scientific evidence while considering this particular test.

In a medical setting, the test for nystagmus is performed utilizing computerized instrumentation called an electronystagmograph (ENG). This instrument gives exact measurements of speed of eye tracking, amplitude and rate of nystagmus oscillation, and the measure of the angle of onset. It also maintains the exact measured distance from the test subject's eyes to the tracked stimulus. These precise measurements leave nothing to guess work and chance and provide the required information to allow interpretation of the results of the study. This should be some indication of the inability of a law enforcement officer in the field to obtain very much useful information from simple observation.

Using ENG for measuring nystagmus, it has been found repeatedly that angle of onset of nystagmus is different in many individuals and occasionally within the same individual. There is one documented study with extensive data that indicates that 50% to 60% of all individuals tested exhibited some form of nystagmus that was indistinguishable from the nystagmus observed in DUI testing, even in subjects with no alcohol in their system. This occurred repeatedly when deviation of the eyes exceeded 40 degrees. Other research documents nystagmus onset at angles less than 40 degrees with blood alcohol levels as low as 0.06%, and at angles of 30 degrees or less with blood alcohol levels as low as 0.048%.

In addition to normal physiological variations, there are other problems that may lead to observed nystagmus that may be unrelated to blood alcohol levels seen in an impaired individual. Circadian rhythms are cyclic changes in animal physiology related to the time of the day in relation to the wake/sleep cycle of the person. These rhythms occur on a regular cycle and on a daily basis, much like the ebb and flow of the tides seen in the tidal plains of the world. Nystagmus is so influenced by the normal circadian rhythms, that it may easily make it nearly impossible for an officer to obtain any useful information from this test that may assist in the decision to arrest or not. Tharp, Moskowitz, and Burns authored an article published in 1981 in the journal *Psychophysiology* titled "Circadian Effects On Alcohol Gaze Nystagmus." In this article the authors recommend law enforcement officers be instructed to adjust their "criteria by about 5 degrees between midnight and 5 a.m." These authors happen to be the same scientist used by NHTSA for their own original studies. Even though the recommendations have been documented, ***NHTSA has never incorporated this recommendation into any of their training manuals or classes.***

Positional alcohol nystagmus has been mentioned previously. This phenomenon is clearly present as the blood alcohol levels are dropping and there is some imbalance between the consistencies of the inner ear fluid to that of the blood. Regression studies have been performed and documented that there is enough nystagmus present from this phenomenon that there is considerable chance of law enforcement officers mistaking this benign nystagmus with the HGN of alcohol impairment. These studies, some commissioned by NHTSA, show results exactly contrary to the teachings in the SFST instruction manuals. One of the peer reviewed articles documented reports of nystagmus being present in some individuals for approximately five hours after all of the alcohol had left the body. This same study shows over half of the participants displayed measurable nystagmus for at least fifty minutes after the blood alcohol concentration measurements reached 0.00%.

A study by NHTSA published in the report *Pilot Test of Selected DWI Detection Procedures for Use at Sobriety Checkpoints* indicated that "at every blood alcohol concentration examined, a much higher percentage of subjects 'failed' the HGN test than showed any indication of impaired driving." It further states that "15 percent of the sober drivers" and "64 percent of those with blood alcohol concentrations from 0.05 to 0.09 percent" failed the HGN test as given by law enforcement officers. This report draws the obvious conclusion that the HGN test will "generally net more innocent victims than inebriates." It is important to remember that this study was a Department of Transportation study commissioned by the National Highway Traffic Safety Administration.

Even after completion of the Standardized Field Sobriety Test training course, many police officers do not fully understand nystagmus, the many different types and causes, and the exact manner in which to perform the test. They do understand the instruction manual and what it has to say about the test; they understand there is a correlation between the observation of nystagmus and a blood alcohol level, but they understand very little else about this extremely complex issue and the actual scientific basis and variations and the potential pitfalls of whatever results they may obtain. Any officer that uses this test for probable cause for making a DUI arrest should be able to testify that they gave the test ***exactly*** in accordance with the directions spelled out in the manual, and the clues, which are quite specific, were present at a certain angle of onset. No police officer should ever be permitted to testify that the test was offered, given, and failed. Allowing the officer to testify to failure eliminates the possibility of showing that the officer may not fully understand the test and clues. The officer should be required to document by testimony what the actual observation were that lead to the arrest.

HGN is considered by many law enforcement officers, attorneys, and judges to be the most accurate and definitive of the Standardized Field Sobriety Tests. It is taught in the training programs to be the most accurate and dependable test available to the officer on the street when attempting to determine blood alcohol content above the impaired legal limit. What makes this such a problematic issue is the fact that there is so little taught about the limitations of HGN, and the many, many different types and causes of nystagmus other than alcohol impairment. No training manual produced today gives any indication to student officers of the significant probability (***not possibility***) of a false positive finding. Even published studies by NHTSA's own scientists indicate the dangers of using this particular test. A test with such a proclivity towards false positives should be viewed with hesitancy at best.

It is easy to see how complex an issue nystagmus can be. As the DOT's admitted "best" test of the standardized field sobriety tests, it creates remarkable doubt toward the validity of the remaining tests in this deeply studied battery. It is quite rare that a law enforcement officer understands the ins and outs of nystagmus when questioned appropriately in court. Even the DOT NHTSA manual has several inconsistencies in the science behind nystagmus. A thorough review of the research into nystagmus and its use as a tool for determining alcohol impairment will be enlightening to anyone willing to devote the time and energy to this matter.

Horizontal Gaze Nystagmus, as a field tool for law enforcement officers must be viewed with concern and caution, and used only with the understanding that there are many possible reasons for a false positive, if it is to be used at all.

Divided Attention Testing

Overview

Driving is a complex activity that requires multiple tasks on a continuous basis the entire time one is to operate a motor vehicle. There is constant visual input, requiring continual physical movement to control the vehicle safely. This is the ultimate divided attention activity. Naturally, throw in a stereo, or even a cell phone and text messaging, and the attention division is much more complex. Top this all off with a drink or two, and the operation of a motor vehicle becomes a much more daunting activity.

When attempting to develop testing to evaluate the potential impairment, the researchers looked to divided attention testing as a valid method of looking at determining if one could safely operate a motor vehicle. The two tests that survived the developmental protocol were the nine step walk and turn, and the one leg stand. Each of these tests has a mental component, memory component, and a physical component, and the clues developed involved all three areas. With a complex instructional phase and the multitasking physical performance phase, it is assumed that a DUI suspect will not be able to complete the test in a satisfactory manner if their blood alcohol level would be at or over a 0.10%. These two tests are clearly unnatural acts, designed for failure from the outset, with the direct purpose of developing probable cause for an arrest for DUI.

The officer on the roadside, when giving these tests, must be aware of the surroundings in order to ensure the fairest possible situation for providing valid results. These tests should be given in an area with a dry, hard, level, non-slip surface. The tests should be given in an area of relative safety in order to eliminate an additional task on the suspect—concern for their own well being. If an area of relative safety and with a proper surface is not readily available on site, it is not unreasonable to move the subject a short distance to an area that would provide the proper test environment. If there is no readily available appropriate test area, the DOT manual recommends these tests not be used, and the officer develop his case with the driving behavior, other observations, and the HGN test.

The mental portion of the tests begins with the instructional phase. The officer will begin with elaborate instructions by initially placing the suspect in a particular stance, and will begin his explanation of the test and the expectations in the upcoming performance phase. The suspect is advised that he must remain in the stance during the instructions and during the demonstration phase until told to begin the test by the officer. The suspect is given detailed instructions on performance of the test and is expected to mentally process the information in order to successfully execute the test procedure.

Once the suspect has advised the officer that he understands the directions and expectations, he is then expected to use the information and proceed to the performance portion of the tests. The performance is graded on compliance with the instructions as given, on balance, and physical coordination. The suspect is expected to perform with absolute recollection of all the instructions, and with perfect balance and coordination.

Failure to stand correctly during the instructional phase, to perform the test in accordance with all the instructional details, and with excellent balance and coordination will provide the officer with the appropriate clues to make the determination of potential alcohol impairment.

Nine Step Walk and Turn

Like the HGN test, the nine step walk and turn is a modification of a common medical diagnostic test used by health care providers for evaluating certain neurological and central nervous system ailments and disorders. When used in its pure medical form, it is not a divided attention test, and is solely a performance based evaluative mechanism to determine a patient's health status. For the purposes of DUI enforcement, the test has been "enhanced" in order to increase the complexity to elicit a different result. Unfortunately, the enhancements have done nothing to screen out the original disorders that the test was designed to diagnose in the beginning.

This test is a modification of one of the oldest DUI determination test that has been used since the earliest days of driving. In the old days of law enforcement, the suspect was made to "walk a straight line" and it was assumed a drunk driver would not be able to do this successfully. The nine step walk and turn is the walking a line with a twist. The suspect not only walks a line, but must do it in a heel to toe manner, with a certain number of steps and an elaborate turn in the middle designed to complicate the test and better judge balance and mental ability to follow instructions.

The officer should always designate a straight line for the suspect to walk for this test. This should be a line that is clear and obvious to the test subject in order to not add in an additional task of "seeing" and imaginary line. There are multiple lines normally in parking areas or on the roadside, with asphalt or concrete seams or traffic lane designation lines. If there is no natural line available, it is not inappropriate to make a line, using masking tape or duct tape if necessary to lay out the proper line.

To administer this test, the officer places the subject in the "start" position, standing erect, and arms at their side, with their feet in the heel to toe position while the verbal instructions are provided and the officer demonstrates the procedure. This actually places the subject in a position with the body held upright upon a very narrow base of both feet in alignment, front to back, under the body. The law enforcement officer then provides the verbal instructions on what is expected for performance of the test, with exact details of how the suspect is to perform the walk, the number of steps, and the turn. After this, the officer will demonstrate the test, having the subject watch through the walk, all the while keeping his feet in the "instructional" position. It is mandatory that the officer complete each of these steps each and every time the test is given, in order to comply with the "standardized" part of the standardized field sobriety test, in order to permit the drawing of any valid conclusion for the results obtained. The manual instructs the officer to complete the instructional phase with haste and with no undue delay in order to prevent the test subject from being forced to maintain the start position for an unusual amount of time, and therefore eliciting a balance failure due to something other than alcohol impairment.

The subject is then asked specifically if he understands all of the instructions and the expectations for successful performance of the test. After advising the officer that he fully understands what is expected of him, the subject is then told to start the test. If at any time the subject does not appear to be able to complete the test in complete safety, if he becomes "leg locked" (legs twisted together and not able to take the next step), steps off the line more than three times, or just cannot perform the test without falling or nearly falling, the officer should terminate the test, and make this observation of the suspects inability to complete the test and why.

When evaluating, or grading, this test, the officer is trained to look for eight specific clues in the performance of the test. The first two clues are checked during the instructional phase of the test only. When the subject is placed in the start position, the subject should be able to maintain excellent balance. Failure to do this is the first clue. The second clue is noted that the subject waits to begin the test when told to do so; should he begin early, during the instructions, this is clue number two. The next four clues are observed during the walking portion of the test—if the subject stops during the walk, misses touching his heel to his toes, steps off the line, or uses his arms to maintain balance by raising them more than six inches from his side are all the next set of clues. The next clue is the turn. The subject should turn in the direction instructed, using the manner instructed, and should maintain balance during the turn. The final clue is the number of steps taken.

Each of these items is a single clue. Should the subject not touch heel to toe on six different steps, you would have a single clue, not six. If he takes more steps or fewer steps out and a different number back (other than the instructed nine), that is a single clue. If a subject raises his arms four times for balance, this is again a single clue.

9 Step Walk and Turn Clues

Failure to Maintain Proper Position During Instructional Phase
Starting Before Being Told to Start
Stopping During Walk Portion of Test
Not Touching Heel to Toe
Steps Off Line
Uses Arms to Balance During Walking Part of Test
Turn Improperly
Incorrect Number of Steps Out or Back
THE PRESENCE OF TWO OR MORE CLUES INDICATES POSSIBLE BAC OF 0.10% OR MORE

Chart E

Interpretation of the performance on this test is observation for these specific clues, and only these specific clues. Other problems or difficulties with the performance of the test have not been included in the standardization of this test, and must be ignored by the officer, and later, the court. If there are two of the eight clues noted, then the assumption is that the blood alcohol level may be at or above 0.10%. If there are no clues or only a single clue, then the implication is that the blood alcohol level may be less. There is no differentiation for lesser blood alcohol levels for lesser performance (such as a possible 0.05% for only a single clue) or for performance deficits in other areas than the designated clues.

When observing only this test (for whatever reason), it should yield an accuracy rate of 68% in determining a blood alcohol level at or above 0.10% in the decision to arrest or not. Research materials indicate that when used properly with the HGN test, the accuracy rate rises to 80%.

When the nine step walk and turn exhibits enough clues for an arrest, but the HGN does not, then the implication is that there may be impairment by some substance other than alcohol, and a DUI by drugs may be indicated.

One Leg Stand

The one leg stand test is again, like the previous two tests, a modification of a standard medical test, the Rhomberg test, used by health care providers to detect serious health conditions and ailments of the central nervous system or peripheral neural pathways. Like the nine step walk and turn, it is not designed to be a divided attention test in the medical setting, and it, too, is complicated by the additional tasks of detailed instructions and physical performance.

As a divided attention test in the performance of DUI evaluation, the subject must stand in a certain position, listen to complicated, detailed instructions before attempting the test, and then perform the test exactly as instructed. Failure to stand properly or follow the instructions constitutes clues to impairment as much as poor balance and performance. By requiring the subject to maintain balance, keep the foot a measured distance off the ground, and count properly divides the subjects attention between the physical acts involved and the mental task of recalling instructions and counting.

The instructional and demonstration position for the subject is to stand erect, with his feet together and hands directly by their sides. The instructions are given for the performance expectations. The subject is advised to raise either foot approximately six inches off the ground with the toes pointing straight out, with both legs kept straight at the knee. The subject is told to look directly at the foot that is held off the ground, and to count out loud, "one thousand one, one thousand two, one thousand three....." until told to stop.

The NHTSA manual ***does not*** instruct the officer to have the subject count to "one thousand thirty" as is normally and frequently seen in courtroom testimony. The officer is advised to time the performance portion of this test for thirty seconds, and should the law enforcement officer have the suspect count to a particular end point, an additional task has been added to the test, and it is no longer a "standardized" test, and therefore, no useful conclusion can be drawn from this test.

As in the other tests, the subject is asked if the instructions are fully understood before being allowed to perform the test. Should he answer that all the instructions were clear and understood, then and only then is he permitted to attempt the test. He must remain in the start position during this entire time, to include arms by the side, even during the physical portion of the test. Like the nine step walk and turn, there are specific clues to observe for evaluating the performance in this test.

The first two clues are observed during the instructional phase as well as the performance phase of this test. First, it is noted if the subject is "swaying" while in the start position. This swaying must be distinct and noticeable in order to be considered a clue. It may be side to side or front to back, but must be clearly not simply natural tremors that may be present from muscle fatigue. The next clue is observation of the use of the arms for balance, again during the instructional and performance phase of the test. Should the arms be raised more than six inches from the sides, the implication is that they are being used to maintain balance. The third and fourth clues are only noted during the performance phase of the test, and are placing the raised foot down onto the ground prematurely or hopping on the foot on which the subject is standing. The officers are instructed to terminate the test at any time should the subject's safety become an issue, or should the subject not be able to maintain balance without falling. The test may be terminated should the subject place the raised foot down more than three times.

One Leg Stand Clues

Swaying Using the Arms for Balance Hopping on One Foot Placing the Foot Down Prematurely THE PRESENCE OF TWO OR MORE CLUES INDICATES POSSIBLE BAC OF 0.10% OR MORE

Chart F

Interpretation is simply the observation of two or more clues. The presence of two of these clues is indicative of a blood alcohol level at or above 0.10%. Should only one clue, or no clues, be exhibited, the implication is that the blood alcohol level is less than 0.10%, and is not any indication of a particular level of impairment. All of the research into this test indicates a reliability level of 65% when used alone to determine an arrest or no arrest decision.

Limitations of SFST

No test known to exist will ever yield perfect results, whether it is scholastic tests, medical testing, SFST, or even evidential blood and breath tests. Many law enforcement officers, prosecutors, and judges are of the impression that these tests, through significant research, have been developed and fine tuned to the point that they will give the correct answer each and every time they are used and interpreted. Like any other test, there are limitations to their usefulness and the ability to draw the proper conclusion when attempting to make the proper arrest or no arrest decision. Anyone, no matter what their position in the system, that is truly interested in the proper enforcement of our DUI laws and punishing only the guilty should not only be aware of these limitations, but should be adamant about understanding these limitations in order to properly apply the information obtained from their use.

Prior to any discussion about the limitations of the SFST, it is essential to understand the test, the research that went into the development of these tests, how they are scored, and how the conclusions are made from the scoring. There are volumes of research documented on these tests, and their development, much of which is government sanctioned as well as private funded. This research explains in great, painful detail, about the tests, how they are to be used, and what limitations there are to their use for determining alcohol impairment. Very few people are willing, or even able, to undertake a thorough review of all of this literature, and therefore, too often, key people in the enforcement of our DUI laws may not fully understand the implications of these tests.

A significant issue in the use of these tests, as in other testing for alcohol impairment, is that there is some uniformity in all people when it comes to physical and mental abilities. If this is true, then all results obtained from performance in any test will fit nicely into a category, either pass or fail. Those that pass are not impaired, and those that do not pass, are impaired. Entirely too often, there is not enough significance placed on the fact that not everyone will fit nicely into one of these two categories. It is true that most people without any significant health issues will be able to perform the standardized field sobriety tests, but there is little tolerance in court for any argument that there are many people that simply have a biological difference or some other logical explanation as to what may have been viewed as poor performance on these tests. Any factor that makes one person different from another may cause difficulty in test performance. Simple changes associated with aging, weight gain or loss, past injury or illness, previous exposures to chemical agents or toxins, or simply variability in genetics may make key changes in ability to perform adequately on these standardized tests.

Human Biological Variability

Variation in one human being to another is probably one of the most important issues when it comes to limitations of any physiological test. The fact that all people are different is so obvious that it is usually overlooked as an issue in this situation. Every person has their strengths and weaknesses—some do well in math, some in history, some are laborers and some are managers. There can be no single standard to measure each person with any level of fairness. Humans are varied by race, sex, ethnic background, genetics, age, and surgical changes. Sometimes, even within a category, there are subtle changes that are nearly impossible to see, such as a person with Turner's syndrome or Klienfelter's Syndrome. These people cannot even be placed into a male or female category with any accuracy as they do not have the same X and Y chromosomes that most of us have. Is it unreasonable to expect all women or all men, all old and all young, all Asian or all Black, or even all American or all Irish to perform the same on a single set of standards? Clearly, not everyone can be measured by the same set of guidelines, particularly in a situation that may remove a person's rights or privileges in a legal setting such as an arrest.

Human variation comes into play every time the SFST are used to evaluate a subject on the roadside. All people vary by age, educational level, physical ability levels, occupational training, and environmental exposures over time and even by exposure to alcohol over time. These variations are not only present from one person to the next, but also can be different in the same person from one day to the next.

Basic human variability causes differences in each person's ability to perform dexterity testing, mental functions, and other physical agility tasks. This variability also alters our responses, including uncontrollable physiological responses, such as heart rate and rhythms, respiratory rates, neurological responses to stress, as well as endocrine responses (adrenaline rush—fight or flight mechanism—in stressful situations). It is difficult for a medical professional in an office setting, much less a law enforcement officer on the street, to measure responses in someone they may have never encountered before this evaluation. Without a baseline evaluation, it could be extremely difficult to know what performance issues are related to biological differences or alcohol impairment.

No two people can be expected to have the same physiological abilities. This will affect how well someone can perform in sports, how well one could sing or dance, and how well one could perform on dexterity tests such as the standardized field sobriety tests. There is little research into specific biological differences from one person to the next when evaluating the clues on these tests.

Biological Influences

Other than biological differences in people, there are other biological issues that must be considered when evaluating performance on these tests. Fatigue changes everyone's ability to perform mental and physical tasks. Circadian rhythms, or changes in biological systems due to time of day, will change responses to stimuli depending on the situation. Diagnosed, as well as undiagnosed, medical conditions can alter the performance when observing these tests. Other chemicals present, whether taken as food, additives, spices, environmental, or even medications (over the counter and prescription) can make significant changes in abilities, sometimes to enhance the ability, and sometimes to limit ability.

It is important to note that most DUI contacts occur late at night, usually after a long day that may have begun as much as eighteen or more hours earlier. A DUI suspect may have already spent a long day of work on the job (physical or mental), taken care of family responsibilities, and then attempted to relax with a few drinks late in the evening. This scenario itself lends credence to the fact that a DUI subject may be suffering from physical and mental fatigue at the time of the contact. Fatigue is nearly as powerful a factor as alcohol may be, and it can significantly alter a person's ability to perform dexterity tests or divided attention tests. It certainly affects the presence of certain types of nystagmus. Muscle fatigue can alter one's ability to perform physical tasks, and mental fatigue may alter the ability to recall fine details of the instructions in the correct manner of giving the standardized field sobriety tests.

Other than the late hour, the normal DUI contact occurs on the roadside. Rarely does a nice smooth, flat, clean surface make itself available for the performance of these tests. Multiple distractions are nearly always going to be present when a subject attempts to perform these tests. It may be difficult to not only recall fine details of instruction, but also to maintain balance while being concerned with debris strewn roadsides, the presence of rodents or other animals, or the nearby passing of other traffic. Even a slope of even a minimal degree can cause a problem that may not be attributed to anything other than impairment.

Passing traffic is an issue for several reasons. The speed and volume of traffic is itself a concern, but the potential strong air currents produced, the flashing of lights of passing vehicles, and the noise present may significantly alter the ability to concentrate on the important issue at hand. Passing headlights and taillights, as well as the flashing lights of the police vehicle (or vehicles), can produce a nearly hypnotic state, or cause a type of mini-seizure that can be misinterpreted as impairment considering the clues. This is uncontrollable and occurs within the central nervous system, and may not manifest itself in any other way, and therefore, will be missed by the officer giving the tests. Flashing lights of this nature are well known to induce central nervous system changes, even in a person with no history of seizure disorder, and in fact, is used to evaluate brain wave function in a common medical test—the EEG, or electroencephalogram.

In today's health conscious environment, it is not unusual for people to take large doses of vitamins and herbal supplements on a regular basis, in the belief that if a little is good, a lot is better. Many of these products directly affect nerve conduction function, particularly in higher doses often seen. Most of these products have not undergone government study and evaluation, as over the counter dietary supplements, and therefore carry no mandated maximum daily dose. These products are sold in health stores, as well actual pharmacies, and may be purchased without benefit of medical expert recommendations of one's health care provider. Even if a law enforcement officer thinks to ask about medications, most people don't consider these supplements to be medications and may not think to mention it when asked.

Nicotine and caffeine are two of the most common drugs encountered on a daily basis. They are normally consumed not just throughout the day, but in higher doses in the same social situations where alcohol may be consumed. Even in subjects that do not smoke, there are frequently measurable levels of nicotine in blood and urine specimens after spending time around smokers (such as bars, restaurants, and homes with one or more smokers). Both agents affect people in a similar manner. Both not only affect nystagmus, but also can cause watery, bloodshot eyes, rapid heart rate, sweating, and poor performance on fine motor dexterity skills. Add to this the fact that both chemicals can act upon the adrenal glands and there can be wide variations in ability to perform the standardized field sobriety tests, particularly if someone has more of either chemical in their system than they are used to.

Medical Issues

In any particular DUI trial, there is a list of symptoms that an officer may produce on the witness stand to indicate a subject was under the influence when stopped. This issues are "flushed face", "poor coordination", "drowsiness", "odor of alcohol on breath", or "bloodshot eyes". Any review of even elementary medical literature for these symptoms will show a host of different medical conditions for each of these observations, many of which are not alcohol intoxication.

Facial flushing is notable in a patient with suboptimal treatment of high blood pressure, menopause, treatment with certain cholesterol lowering agents (including common over the counter niacin), and diabetes mellitus. The *odor of alcohol* on the breath could be very easily mistaken for acetone in the breath. This is commonly seen in poorly treated or even as yet undiagnosed, diabetes, seizure disorder, bowel problems (including blockage or decreased motility), or head injury. A person on a tight calorie restricted diet will burn fat instead of the normal sugar obtained from the regular diet, and this will give a ketone, or alcohol-like, odor to the breath. *Poor coordination* may be seen in a subject with muscle spasms or injury, shock or exposure to some industrial chemicals, as well as some household cleaners or gardening chemicals. Someone may appear *drowsy* due to anemia, mild viral illnesses (even if they have not yet otherwise made their appearance known), or certain over the counter or prescription medications. It should be obvious that most of these situations will most likely have little or no effect in one's ability to operate a motor vehicle safely, yet can clearly be used to persuade a court that a driver may be impaired.

There are certain metabolic illnesses and disorders that may influence a subject's ability to perform the SFST as instructed and expected. Diabetes mellitus, the many forms of arthritis, and neurological disorders may cause significant difficulty in performing adequately. These disorders and illnesses are not rare by any means, and are so common that it is almost definite that all officers will come into contact with subjects that suffer from these problems on a regular basis. It is not unusual for a single subject to actually have more than one of these problems, as many of them go hand in hand together, and some may not even have been diagnosed for some time after initial onset.

Diabetes mellitus is a major health issue and is one of the most common health concerns to our society today. It is a condition all too common in our society, particularly in middle aged and older subjects. Normal consequences and complications of diabetes are nerve damage, blood vessel damage, and electrolyte imbalances (affects muscular control). The earliest complication of diabetes is peripheral neuropathy, or the breakdown of the nerves to the extremities. Chronic low grade pain in the extremities may be present for years before the issue becomes significant enough to seek care and damage is normally already done.

Arthritis is a generic term indicating joint inflammation and pain. When this occurs in the joints of the lower extremities, it may drastically reduce the ability to perform the balance and standing portions of the standardized field sobriety tests. This, like diabetes, accounts for a major number of medical office visits each day, in both the specialist and primary care arena. With many years of upright walking, running, and jumping, the wear and tear on the major joints of the lower extremities takes its toll. Joint replacement surgery, from hips and knees to toes, is becoming a huge portion of most orthopedic surgical practices. Healthy lower extremity joints are paramount to the ability to balance and perform the complex maneuvers of the SFST.

Osteoarthritis is simple wear and tear on joints. They simply wear out, either in the articular surfaces of the bones, or the cartilage lining the joint spaces. Rheumatoid arthritis is more insidious and is an eating away of the components of the joint in a manner similar to one being allergic to their own bodies. With this type of arthritis, it is not only the joint surfaces themselves that suffer, but the ligaments, tendons, and membranes responsible for lubricating the joints also. Pain and instability will certainly produce positive clues in the SFST.

Fibromyalgia and fibromyositis are two very similar conditions, and, like arthritis, are chronic pain conditions that are becoming more and more common in our society. Unlike arthritis, these conditions originate in the muscles and other connective tissues (tendons and ligaments) of the skeletal system and not in the actual joints. Both of these conditions are extremely difficult to diagnose, and in fact, until recently, have been considered more of a mental condition than physical. The medical community now knows better. There is very little to treat these disorders, and almost no diagnostic tests, so it frequently goes undiagnosed and untreated. Due to chronic muscle pain and instability, the implications to the SFST are obvious.

Neurological disorders may be located in the brain, the spinal cord, or in the peripheral nerves. Migraine headaches are one of the most common neurological disorders in modern medicine. Although they usually manifest themselves as a severe headache, this is only part of the disorder. Most migraines present with what is known as an aura, or pre-headache condition, that may last anywhere from a few minutes to a few hours. This may consist of visual disturbances, including color vision changes, visual field deficits or tunnel vision, stinging and watering of one or both eyes, and even a temporary complete loss of vision to one eye. Some migraine sufferers may never even get the classic headache, but only the aura. It is possible that many migraine sufferers may never seek medical care for fear of being labeled a drug seeker or due to the stigma of going to the doctor for "just a headache."

Nystagmus may be drastically enhanced by the aural and optical prodrome to a migraine. Pupillary reaction may be drastically altered even to the point of one pupil appearing blown, or much larger than the other. The ability of a person to concentrate to any detail may be significantly hindered by the presence of a migraine, either from the pain, the nausea, or the aura. Vision, balance, and other fine motor skills may be affected to the degree that heel to toe walking may be difficult, the turn may be nearly impossible to perform with any grace, and the inability to concentration on details will eliminate any appearance of lack of impairment. Severe nausea and vomiting is a common presentation. Positive clues will certainly be present, even though they may be coming from the migraine and not alcohol.

Cerebrovascular disease, or strokes, can be present in many different degrees. There may be something as minor as transient ischemic attacks (or TIA's) to something as severe as blocked or blown out arteries in the brain that cause immediate paralysis or sudden death. TIA's may go undiagnosed for many years in some subjects as the symptoms could well be noticed only as mild nuisances that last anywhere from a few seconds to an hour or two. The fact that they are transient is what makes them minor, not the fact that the symptoms are not significant.

The symptoms of TIA can be exactly the same as the symptoms of major stroke. There can be slurred speech, nystagmus, partial paralysis (either to small groups of muscles or entire limbs), mental confusion, and inability to perform fine and gross motor skills. Major strokes are generally a very serious condition, and in the DUI context may be noted as a major past medical condition. Patients with previous history of major stroke may exhibit considerable signs of impairment, including abnormal gait, a slapping of the feet on the ground, a limp, and the inability to stand on one foot. Occasionally answers to what appear to be very simple questions may seem out of context. Recovery from major stroke can be what appears to complete on the outside until one starts looking closely for minute signs of impairment. There could be something as simple as slurred speech or wavering when standing that may appear to be related to poor balance possibly with the look of impairment. These signs may be more noticeable on one day more so than on another. The implications here for DUI enforcement are staggering.

There are also neurological problems that may be present from other conditions than the obvious stroke or TIA. Past trauma, even minimal, can cause problems with discs in the vertebral column. When disc injury occurs, should there be a herniation, there may be pressure placed upon the spinal nerve outlet from the spinal cord. At what ever level this herniation may occur will dictate what outward sign of neurological deficit may be apparent. Should it be present in the lumbar area, then the deficit may potentially make it very difficult to perform any maneuver that may require exact placement of the feet or the holding of one foot off the ground. Should the injury be in the cervical area, there may be some deficit to the movement of the facial muscles, and may be misinterpreted as a sign of impairment with speech or tracking of the eyes. Not all disc injuries are significant enough to warrant medical evaluation even thought they may be present to the degree to cause minimally visible signs to the observer.

Multiple sclerosis (MS) is becoming a more frequently seen and diagnosed illness in today's medical community. It normally affects women, and normally in their third and forth decades, but may be so insidious in its presentation so as to go undiagnosed for many years. MS is a demyelinating disease, which is a disease process by which the myelin sheath of the nerves begins to break down. This myelin sheath is actually the insulation to the nerve, much like the insulation on a wire in a circuit within a building or automobile. As the myelin breaks down, the nerve itself begins to "short circuit" just as a wire in an electrical circuit would do. When this happens, the signal obviously does not reach the site intended by the brain, and whatever action expected cannot occur. These neurological dysfunctions have obvious implications in DUI enforcement as there can be apparent inabilities to perform several aspects of the tests which may easily be mistaken for impairment.

Accidents and Injuries

The human body has an amazing ability to protect itself from blood loss and pain by normal physiological responses to trauma. Frequently, a subject may be injured in an accident, yet at the scene deny any problem, and thereby not be evaluated by rescue personnel or transported to the emergency department of a hospital. It is not uncommon in a medical practice or ER setting to have a patient present three to five days after an accident with pain, and yet, not have been treated at the hospital immediately after the accident.

When an accident precipitates a DUI investigation, the driver of a vehicle under suspicion could have potential injuries that are unknown to the subject or the officer at the scene. The obvious injuries of fractures, major bruising, lacerations, and head injury will certainly have immediate implications at the scene for DUI evaluations. No officer would consider requesting a suspect perform SFST with these obvious injuries, although it is unlikely (appropriately so) that they would forego the SFST when there is no obvious sign of injury, no matter how violent the crash.

There may be potentially devastating injuries to the lower extremities resulting from a crash that may not be obvious to the investigating officer or the suspect. It is well documented in medical literature about subjects with fractures of the hip or ankle being able to ambulate without difficulty after a major trauma. Fear, adrenaline, shock and other issues may well mask a person's ability to feel pain for some time after a major traumatic event, even though they may perform poorly on SFST due to these injuries. Poor performance on SFST could easily be attributed to impairment by the investigating officer instead of undiagnosed or unnoticed injury immediately after an accident. Not all fractures or concussions will cause pain when adrenalin is rushing through the body.

Head injuries may be much more insidious and difficult to detect after a violent crash. An injury serious enough to cause a subdural hematoma may not be noticed by symptoms for up to twenty four hours or more after the injury. There will begin to be immediate subtle neurological changes that may go unnoticed by the suspect, but may appear to be signs of impairment to the well trained, experienced investigating officer. There may be nothing more than a minor headache, which may not be significant enough to prompt the call for an ambulance. Head injury could still be significant to the degree that would limit the ability to not only perform the SFST, but also to understand the directions enough to even begin to perform.

Nystagmus ***should never be used as an indication of impairment in any subject that has suffered a head injury***, no matter how minor. Closed head trauma will affect the ability of the eyes to track smoothly for at least several hours after the event as trauma to the surface of the brain, no matter how minor, may cause significant enough changes to the ability of the eyes to track even with the total lack of other symptoms. Care should always be used when evaluating any subject after a motor vehicle accident no matter how minor, as even a minimal impact can cause the head to strike the A-pillar, the steering wheel, side window, roof, or windshield of the vehicle. Should the vehicle's air bags deploy, there will definitely be enough facial trauma to cause interference with the observations of nystagmus. Newer vehicles with side air bag curtains add another twist to the equation, as deployment of these protective devices would strike the side of the head with some degree of force.

Any level of jerking trauma from sudden deceleration in a motor vehicle crash will certainly throw the inner ear fluid into disequilibrium and cause difficulty with balance, as well as tracking of the eyes. This force could be a front to back jarring from a frontal or rear end collision or the lateral twisting force of a side impact strike. Any of these sudden head movements can set the inner ear fluid in motion to a significant level to cause difficulty in assessing the SFST results.

With the human body's ability to protect itself after trauma, even some internal injury with bleeding can be limited by normal anatomy (splenic capsule) for several hours after a traumatic event. Bleeding may be significant enough to cause difficulty with balance, but not significant enough to cause a major amount of pain for several hours. Any motor vehicle accident warrants care with interpreting the clues obtained from all of the SFST in all cases in order to ensure accurate decisions to arrest or not to arrest.

Environmental Effects

It is difficult to assess what chemicals may be involved in the behavior and ability of a subject to perform on these tests. There could be occupational and leisure activity exposures to solvents, paints, insecticides, fertilizers, gases, and petroleums of all types. Short term exposures could cause mild and transient difficulties, where long term exposures can cause neuropathy similar to that seen in many chronic disease states, such as diabetes or stroke. Either of these exposures could cause difficulty in performance on the standardized tests.

Certain environmental conditions can affect performance on the standardized field sobriety tests, and the interpretation of the results observed by the investigating officer. Abnormal temperatures, either hot or cold, precipitation and other weather factors, and even clothing can come into play in the ability to perform field sobriety tests.

Cold temperatures can cause constriction of superficial vessels, and may cause blanching of the skin. Blood shunted to the inner areas of the body can cause less than ideal perfusion of the central nervous system. Muscles can be affected by cold temperatures, and sudden changes from warmth of the heated vehicle to the cold of a winter night. Hot summer temperatures can cause hyperthermia, or elevated body temperature, that can affect muscle control, alter electrolyte balance, and cause rapid heart rate and respiratory rate. Sweating can cause an issue as minor as a nuisance, and as severe as burning and redness of the eyes, as well as affect nystagmus.

Rain, sleet, or falling snow can be a major distraction when attempting to perform the standardized field sobriety tests. Standing water, ground covered in snow, or icy patches on the ground can have obvious consequences. Where the officer may be dressed for outdoor exposure in foul weather, the subject may be inappropriately dressed due to plans for being indoors with minimal outdoor exposure. Even on the warmest of summer nights, wet hair and clothing can cause a chill to the body by evaporative cooling of the skin. This may cause shivering and other difficulties with attention to detail and fine motor control. Lightning flashes can have the same affect on a subject as the flashing of passing vehicle lights or the flashing of emergency lights on the police vehicle.

When a person sets out to travel to a particular location, it is likely that they will be dressed for the occasion to which they are traveling, and may not be dressed appropriately for a stop on the side of the road for a potential DUI investigation. Expectations for performance on field sobriety tests for someone significantly underdressed for inclement weather conditions should be adjusted accordingly. Footwear is most likely to be the biggest offender in inappropriate clothing for the occasion of performing standardized testing. Often, shoes may be ill fitting (but fashionable!), could be high heeled, or pointed toed, such as some women's dress shoes or men's boots. Soles could be so slick as to make it very, very difficult to perform to any level of satisfaction in the two divided attention tests. Loose fitting sandals may cause other problems, particularly when attempting to stand on one foot, or walk heel to toe. Bare feet, frequently an option selected when other available footwear is less than ideal, is not a suitable choice in an area on the roadside that may be covered in trash, gravel, sand, oil, or broken glass. It would be nearly impossible to concentrate on the task at hand when one is concerned about pain and injury to the feet.

Physical Limitations

Advance age and obesity are the two primary restrictions that have warnings within the NHTSA instruction manual. The manual specifically warns that any subject over the age of 65, or that is more than fifty pounds above their ideal body weight should not be given the nine step walk and turn test or the one leg stand test.

Physical agility and abilities are degraded as we age, and all research into the SFST showed that on average, most people over the age of 65 had diminished ability to the degree that would interfere with their normal ability to perform the SFST. Nerves conduct information and instructions throughout the body. This is done much the same as in electrical circuits in our homes. Nerve impulses are actually a function of movement of negative and positive ions, exactly as electricity moves through wires in a circuit. Just as aged wiring in older homes needs to be replaced due to damage from years of use, nerve function also deteriorates as we age. Insulation breakdown may allow some short circuiting, and aged nerve cells may cause increased resistance. Either way, the message may not get to the appropriate muscle groups in order to permit appropriate performance on these tests.

Obese conditions make balance and physical agility very difficult. A person with fifty extra pounds of body weight are very likely not physically fit, and standing with one leg off the ground for thirty seconds may be a major task. Fifty extra pounds of body mass may make it totally impossible to even see the feet in order to assist in placing them heel to toe as they perform this walking test. Multiple configurations of body mass in obese subjects cause significant issues with center of gravity in the obese.

Not every subject over 65 will be in less than ideal physical condition, and it is certainly true that even some subjects much older may be able to perform well. The same may be true for some subjects more than fifty pounds over their ideal body weight, but the DOT, in the NHTSA training manual, has set a limit that anyone in these two categories should never be given the nine step walk and turn or the one leg stand tests. This guideline is based in sound medical science, and should always be followed.

Human beings, like all other mammals, are trainable. Muscle memory is a powerful mechanism. This is made obvious by the fact that law enforcement officers must train for hours with their duty weapon in order that should the unfortunate occasion arise to use the weapon, minimal thinking will have to go into place to draw the weapon, bring it on target, and fire if necessary. This is muscle memory at its best, and can be trained into actions that may be made second nature.

The same function happens with the standardized field sobriety testing. This may fall under the heading of "practice makes perfect". Law enforcement officers need to remember that they are well practiced in performance of the standardized field sobriety tests, from the training classes as well as repeated demonstrating when instructing the suspects on the street in the performance of their DUI investigations. What appears to be simple and second nature to the officers and courts may be very difficult for someone that has never attempted to perform these unnatural physical acts. Performing poorly on these tasks that are not normal adult activities may be more natural even for someone with good physical abilities.

One attorney was overheard in court asking an officer why he felt the defendant did poorly on these "unnatural physical acts". Even though this may seem like a comical question at first, it is true that these tests are in fact unnatural physical acts. None of these tasks involve normal functions and activities most people would perform in any other setting. Normal, everyday citizens would have no reason to have practiced these acts, where the law enforcement officer on the side of the road likely has performed this test hundreds of times.

Officers taking a Breath Testing Operators course, as part of that course, will normally consume some alcohol and take the breath test as part of their training. During their training, some officers have been given the SFST and have been found to do quite well, even though they may have blood alcohol levels in the range that would normally give clues for impairment. This comes from the practiced performance of these tests over the years of making DUI inquiries. Human muscle memory is very powerful, and when a task is performed over and over, it becomes quite easy to perform well.

In light of this, a person asked to perform the standardized field sobriety tests should be evaluated in light of their occupation and physical abilities in order to assist in drawing the proper conclusions. Just as there may be false positives and misleading results, there could also be false negatives, and an impaired driver may be released to continue being a public danger.

Validation of SFST Battery Below 0.10%

There has been a nationwide move to lower the legal presumptive limit of alcohol impairment from the 0.10% to the level of 0.08%. When all the original research was performed on SFSTs, it was necessary to standardize these tests to 0.10%, but with the newer, lower limits, NHTSA has made an effort to allow conclusions that would enable detecting drivers with a blood alcohol level of 0.08%.

In August, 1998, a new report was provided by NHTSA that detailed the newest research into standardizing the SFSTs to the new 0.08% level. This report is beginning to show up in court on a regular basis, being used by prosecutors and law enforcement officers to "prove" that the field sobriety test battery is effective at predicting blood alcohol levels below the 0.10% limit. It is imperative to understand the scientific limitations of this study in order to prevent using this report inappropriately. It may be more important to know what is *not* printed in this report than what is printed, in order to not simply accept this report at face value. Not everything is always as it appears on the surface.

It is necessary to start this discussion with the basic idea that all the original research into standardizing the field sobriety tests was directed at determining 0.10% blood alcohol levels. All the research clearly demonstrated that the tests, when used exactly as developed and prescribed, was accurate at detecting this level of impairment. Officer training manuals and programs made it clear that there were certain limitations in the use of this test battery, one of which was ***that no conclusion could be drawn about blood alcohol levels below 0.10%.***

When it became clear that more research was going to be required to re-standardize these tests to a new 0.08%, the NHTSA went looking for a police agency to use for this program. There was a specific criterion that the government had in mind when looking for this department. The agency needed to be in a jurisdiction with the legal limit of 0.08% for arrest, have a zero tolerance policy to drinking below the age of 21, have the ability to generate large numbers of DUI investigative stops, must have a large number of officers already trained in and using the current standardized field sobriety tests (from a certified NHTSA instructor), and have a disproportionate number of younger drivers to ensure a wide range of alcohol levels in suspected impaired drivers. The study needed a large number of investigative stops in a short period of time in order to not prolong the program and still provide a large number of data points for evaluation.

The city of San Diego, California police department had a specialized traffic squad tasked with making DUI arrests as their primary duty function. This dedicated DUI enforcement squad operated from within the regular traffic squad, but consisted of officers very highly trained in detecting and evaluating impaired drivers. This geographic area is home to several very large US Navy bases, as well as the US Marine Corps and US Navy west coast based recruit training commands, and several very large Universities. The command levels of the police department, as well as the prosecutor's office agreed to allow a variation from normal procedures to allow only the three standardized field sobriety test to be used for evaluative purposes, and the officers decisions to arrest or not were to be based solely on the performance of these three tests and no other field evaluation maneuvers.

New guidelines were determined, and the officers of the DUI Enforcement Squad were retrained in the new guidelines. After observing the performance in the standardized field sobriety tests, the officers were to make a prediction of blood alcohol levels, and all appropriate stops were to result in an evidential breath test in order to compare the predicted level with the actual level on the evidential breath test. All the information was transferred to a special form and passed to researchers for evaluation and integration into usable data.

These officers generated 297 usable traffic stop encounter forms for data in this study. Of those, sixty five cases resulted in warnings with an average estimated BAC of 0.060%, but with an actual measured average BAC of 0.044%. There were fifteen additional cases where citations were issued for non-alcohol related offenses, with an average estimated BAC of 0.055%, and an actual measured average BAC of 0.046%. Of the 217 cases that resulted in arrest for DUI, the average estimated BAC was 0.138%, with a average measured BAC of 0.150%.

This data set shows the officers were very careful in making DUI arrests, and actually underestimated BAC at the higher levels, and overestimated BAC at the lower levels. When all the data from this study is combined, there is an average estimated BAC of 0.117% against the average measured BAC of 0.122%. The report indicates this is an excellent correlation between estimated and measured BAC on behalf of these highly trained officers. This fact is true, but is only true when looking at all of the data generated. This may not be the most accurate way to view this data, since the study was designed to evaluate the SFST in detecting blood alcohol levels below 0.10%.

The most important cases in this study may be those where the officers made the incorrect decision to arrest, and the resulting BAC was below the level of 0.08%. There were twenty four cases that fit this criterion. In those twenty four cases, the average predicted BAC was 0.092%, although the average measured BAC was only 0.057%. Variations in estimated and measured BAC were 0.08% estimated to 0.05% measured, 0.11% estimated to 0.05% measured, 0.08% estimated to 0.02% measured, and 0.12% estimated to 0.043% measured. In sixteen of these cases, the clues noted in the HGN tests were consistent with the estimated BAC, but it is obvious the correlation was quite inaccurate.

These officers presented researchers with data that they predicted BAC levels below 0.04% in twenty nine traffic stops, but subsequent testing showed only fifteen of those estimates were accurate. This is a 52% accuracy rate at this low level, yet this report from NHTSA states that the data indicates "...Standardized Field Sobriety Test Battery accurately and reliably assists officers in making DWI arrest decisions at 0.08 percent BAC."[5]

There are several issues in this study, from the design as well as the data and conclusions drawn, that must be addressed before the study should be accepted as proof of ability to determine BAC below the level of 0.10%. Certainly this report highlights all the positive findings in the data. This is in no way an attempt to reduce the significance of these findings. As in any study, the negative findings are also very important, and tell a story that needs to be addressed also. The protocol needs to be evaluated in order to determine how much weight is given to the results and the conclusions the report wishes to clarify.

This study, although there may be some disagreement with the final conclusion, was well thought out, and was set up in a manner to best evaluate this information. There is no attempt here to discredit the scientists that set up and performed this study or their results and conclusions. This is simply an argument about other possible conclusions for the same data and a scientific reason to not permit this report to be used in court as evidence of the validity of the standardized field sobriety tests.

The original research into development of the standardized field sobriety tests was much more extensive than this newest research study, and was directly charged with determining BACs at or above 0.10%. A review of the multiple reports generated from this original research indicates a reliability rate of 83% for detecting impaired drivers at or above 0.10%, and this was only under ideal lab conditions and not imperfect field conditions. This new research gives a reliability rate of 91% (an even higher 94% if you throw out the ten "false positive" results), in only field conditions, to determine a narrower, more precise end point.

In light of this information, it should be asked how these tests can be more reliable at a tougher standard and lower level, than the test's reliability level were at the original 0.10%. The SFST were essentially unchanged for the new evaluation, yet they are being represented to be even more reliable and sensitive than before.

If this question is raised, then the study should be scrutinized to make an attempt to determine where there may be a reason for this more accurate finding. One very likely reason may be the fact that the training and experience level of the officers used in the new study is much greater than the level of even a regular patrol officer, as used in the earlier studies. In fact, not even the entire DUI Enforcement Squad of the San Diego Police Department was used, as some of the officers didn't have the experience the researchers wanted in order to be included in data acquisition. Obviously, the more experienced and more highly trained an officer is, the more likely he will be to make the best decisions possible for determining alcohol impairment.

The DUI Enforcement Squad of the San Diego Police Department is one of the most highly trained and effective law enforcement units in the United States at detecting and arresting impaired drivers. This squad has been in existence since at least 1977, and their entire purpose in being on the road is to locate, stop, and arrest impaired drivers. They perform this function every single day, every single shift, and through sheer repetition become very good at what they are doing. These officers may experience in one month of duty a DUI volume that a normal patrol officer may experience in an entire year, or in some cases, an entire career. This newest report states "The researchers also found that inter-rater reliability increased in subsequent sessions, indicating the important role of training and experience in achieving accuracy, reliability, and overall proficiency."[6]

Even with their extensive training, experience, and above average ability, there were still a significant number of problems in predicting blood alcohol levels below 0.10% as indicated in the above explanations. At levels above 0.10%, the officers indicated that they expected a high degree of accuracy in predicting blood alcohol levels, but, in fact, even these experienced officers tended to slightly underestimate blood alcohol levels (not a bad trait in a DUI enforcement officer—it would always be best to underestimate than overestimate). In the blood alcohol level range the study was designed to evaluate, there were twenty four cases in which the levels were over estimated ***significantly***.

The police officers selected to perform this study were highly trained, highly experienced officers. Each had at least three years of experience as a regular traffic officer before being selected to the specialized DUI enforcement unit. Each is highly trained and experienced in the use of the original standardized field sobriety tests in order to make the appropriate arrest/no arrest decisions. The use of these highly specialized officers was both necessary for the study to eliminate as many variables as possible, and was a detriment to the study as they are certainly not representative of regular police officers in training and experience that are going to be using these tests on a daily basis.

It must be argued that caution should be used in the application of these field sobriety tests, particularly in light of the overestimation of blood alcohol levels at the lower levels experienced in this most recent study. When this study is produced in court in order to prove the accuracy and validity of these tests, the strongest objection should be made as the information is certainly hearsay at best. If this is unsuccessful, then the limitations to this study need to be addressed as clearly as possible. With any scientific study, all of the data and all of the information must be reviewed in order to make the appropriate determination. To read only the conclusion portion of a scientific study, and use that as evidence of any argument, is a mistake made entirely too often. All the data and information must be presented in full in order to tell "the rest of the story!"

SFST Conclusion

The Standardized Field Sobriety Tests may seem like a simple set of tasks on the surface, easy to perform, and giving the ability of the officers on the street to easily determine the impaired status of drivers, but they are an extremely complex set of tasks that have been studied and tweaked in order to provide law enforcement with a tool to perform a very important function in public safety.

There is so much involved in the administration of the tests on the roadside, and the performance of the tests that it is a very difficult situation for everyone involved. Great pains have been undertaken to develop an excellent set of tests that have the best opportunity to provide the most information, with the least amount of danger to the subject being evaluated and the officer administering the test. Certainly the best possible battery of tests has been developed with years of research to back up the manner in which the test is used. These tests are so complex that it is apparent that there are some principle players in the area of these tests that simply do not have a full understanding of their use and their full abilities, as well as their many limitations.

There are limitations in the use of the Standardized Field Sobriety Tests Battery that cannot be removed from the real world work environment of the every day street cop. The tests are given in real world situations that are never ideal. The training and experience of using the SFST are but a small part of most officers regular duty activities, and due to other responsibilities, sometimes take a back seat to more important duties. Physical and environmental conditions are what they may be, and cannot be changed. The officer on the street must deal with what is available, and make the best of it at the time.

No police officer ever wants to convict an innocent person of any crime, but it must be obvious that the officer has already made a determination that the suspect is guilty of the crime of driving under the influence, or they would not have made the arrest. There is, because of this, a certain amount of bias on behalf of the officer in observing the performance of the field sobriety tests. That bias is human nature, and is totally unavoidable. Thankfully, the human traits of all police officers cannot be removed, nor would we want officers on the job that didn't have that humanness.

The Standardized Field Sobriety Test Battery is not perfect. There are certainly conclusions that can be drawn from performance on these tests, and these conclusions are valid only under certain circumstances. The prime premise is that the test ***must be administered exactly as prescribed in the training manuals*** developed from the initial research into these tests. There is ***absolutely no room for any variation from the prescribed manner in which to use and interpret these tests.*** It is clear from all research that these tests must be standardized for all officers regardless of training levels, experience, and geographic location in order to allow any useful information to be obtained from the test procedures.

If these guidelines are followed, and the tests are administered properly and scored properly, then the opportunity to make the appropriate decision is increased. It may not be perfect, but nothing in the world of law enforcement usually is. It is a tool and just like any other investigative tool, it must be used in conjunction with every other tool available in an officer's arsenal.

When it comes to getting impaired and dangerous drivers off the street, it is in the best interest of public safety to make the best decision possible, but this must always be balanced by the need to ensure that the rights of the accused are not trampled upon. Proof beyond a reasonable doubt is still the standard by which a conviction must be measured.

Chapter Six

Use of an Expert

When an issue arises that may appear to have scientific merit it would be wise, and probably mandatory, that you consult an expert in order to properly determine the validity of any possible scientific defense and the appropriate use of that defense. It would be inappropriate to ignore a potential valid defense in any matter. In light of the cost of consulting and retaining an expert, it is no easy matter for any client, but a client should certainly be made aware of the fact that there may be a potential scientific defense. But how do you determine the proper expert when looking at a particular case?

There are several different types of forensic experts available for consultation in these types of cases. The most common type of expert is the forensic toxicologist, and this is the type of scientist most commonly employed in the state crime labs and, therefore, are the most common expert to be encountered as rebuttal witnesses to anyone the defense may bring in. Another type of expert is the medical doctor that specializes in forensic medicine and normally performs an analysis on living or dead subjects, and can provide expert opinions on their particular areas of expertise. A rarely encountered expert is the forensic pharmacologist and this can be a pharmacist or pharmacological doctor (Pharm D.) that has studied the science of medications and their applications to the laws. An even more infrequently encountered expert is the forensic physiologist, a scientist with the expertise of the functioning of the living systems themselves. This can usually involve how the systems interact with different chemicals and the processes involved in testing and evaluating a client for impairment.

Each of these scientists is a potential expert in the field of alcohol and the metabolism of alcohol and how it may affect the body. They all have different training and backgrounds, but all are similar to the degree that they may have an area of interest and expertise in alcohol and how it affects the body. Each has their strengths, and each has their weaknesses for any one particular issue that may need exploring in a particular matter or trial. Not every one of these experts will be able to provide an opinion in every matter, or even involving different issues in each particular matter. There well could be a case in which more than one of these experts may be required to expound on different facts and issues in the same trial.

Any scientist that provides and evaluates information about their particular area of expertise and how it pertains to its application in the law and legal issues, may be considered a forensic scientist. Forensic science is simply defined as the science as it applies to the law. Although there are now training programs in forensic science, and there are even college and graduate degree programs in forensic science as well as specialized areas of forensic science, any scientist can have training in their field of science as well as training or experience in the law, and can easily be considered a forensic scientist. The important issue is the level of training and experience that would qualify that scientist to apply their particular field to the law. Each of these scientists should never be expected to, nor should they be willing to, provide any legal opinion. They can be expected to do no more than explain their area of expertise and how it may affect a particular case at hand. The legal arguments are then up to the attorneys, and of course, the court. It takes an attorney who is well versed in the use of a scientific expert to tie the scientific opinion elicited into the legal parameters required in the matter being tried. It is not the responsibility of the scientist to render any legal opinion.

Many scientists function throughout their entire career never setting foot in a courtroom and never being asked for a forensic opinion. It takes a specialized scientist willing to go beyond their particular field of study to be willing to enter the legal arena and undergo the scrutiny and occasionally demeaning world of courtroom testimony and occasional theatrics. There are frequently opposing experts that will offer differing opinions, and it needs to be understood that even though they may appear to be at odds, there could be reasonable explanations as to the superficial differences of opinions. At no time should any forensic expert be an advocate for the defense or the prosecution. They should always be on the side of justice and truth from the scientific standpoint, and should clearly be interested only in assisting the trier of fact, be it judge or jury, to understand the scientific issue and apply that understanding to arrive at the appropriate verdict. At no time should an expert be retained and paid depending on the outcome of any trial, as the outcome should be of no real concern to the expert. The only issue that should be of interest to any forensic scientist is that the explanation be unbiased and accurate within a reasonable degree of scientific certainty. It is virtually impossible to state any scientific fact as hard and fast, and it should be understood and made clear that the matter is investigated using the available information, most always after the fact.

Each type of potential forensic scientist has their particular weaknesses and strengths. Any matter up for scientific evaluation and explanation needs to be thoroughly investigated in order to determine the best scientist for the issue at hand. The best scientist for a particular case must be determined on an individual case basis and should be evaluated to determine who the best possible expert may be to deal with the facts of that case. There have been cases in the past where the expert hired was not determined to be appropriate, and the opinion was not accepted by the court. This issue needs to be clearly determined before trial in order to avoid any impropriety at the time of trial.

Toxicology is the science involving the study of toxins, or poisons, and how they affect the body. Most toxicologists are laboratory scientists that perform the analysis that is presented to the courts as evidence in a trial that there is a substance present in the body and how much of that substance is present. Generally, there are predetermined limits as to what level of a toxin is dangerous, either from an impairment level, or a level that may cause injury or death. These levels are set by research, or in some instances, such as with alcohol, by statute from the state legislature. These scientists may have an education level anywhere from a bachelor's degree, through a master's degree, a doctoral degree, or post-doctoral training. They are highly qualified to provide information about the testing of different body fluids or tissues, and how the results were obtained. They may have some ability to expand their information about how the compounds may have affected a subject, depending upon their additional training and study on effects of toxins on the living systems. A large part of the toxicologists training involves effects of the toxins on a subject, and therefore it is not out of the realm of their expertise to provide information on levels of impairment.

Medical doctors that specialize in forensic medicine frequently serve as coroners and medical examiners for government agencies, be it local, state, or federal levels. These doctors may be called upon to appear on the scene of crimes to render a medical opinion as to cause of injury or death, and then further investigate by autopsy, physical examination, or fluid or tissue analysis to provide evidence to support that opinion. Frequently, they become involved only when the body or tissues are presented after the fact for evaluation. A physician that specializes in forensic medicine may see live patients in the clinical setting for evaluations and determination of scientific merit of medicolegal issues when that client presents for evaluation after some interaction with the legal system, be it criminal or civil.

Rarely would these physicians perform the analysis of the tissues and fluids themselves, but they would rely upon the forensic toxicologists in the labs to perform those analyses and provide a certified report to complete an investigation. Both of these scientists may then be called upon to appear in court to explain their findings, although they both have very different roles. There are occasional medical doctors that specialize in much narrower fields of forensic medicine and would only be valid witnesses in their particular area of expertise. These could include ophthalmologists, orthopedists, gynecologists, geneticists, or a host of other medical specialties and each of these would be utilized only in their respective fields of specialization. Naturally, it would be impossible to utilize a forensic gynecologist in a case involving an issue with an eye injury. Not all specialists are able, and even fewer are willing, to be used as a forensic expert in a case, even when that case involves their own patient. Most medical doctors are quite uncomfortable in a courtroom, and have significant concerns about confidentiality of their patient's health information, as well as their willingness to spend the required time away from patient care in order to appear in court at a trial. It could be a grave mistake to bring in a medical doctor to provide an opinion in a matter when that doctor is uncomfortable or unwilling to appear on behalf of a client. In this type of case, it would be necessary to obtain certified copies of medical evaluation records, diagnostic testing, and treatment plans and have a separate forensic health care provider review these records, possibly interview the client and physician, and provide an opinion in writing and/or in court at trial.

A pharmacologist could be qualified by education and training as a pharmacist or a clinical pharmacologist. Most clinical pharmacologists are pharmacist with advanced training in clinical medicine and the physiology of the effects of drugs or medications on the body. All pharmacists are trained in extensive programs leading to state licensure to practice the art and science of pharmacy. This involves the act of filling prescriptions, as well as counseling patients about health issues and drug interactions and side effects. A good pharmacist is a highly regarded asset to the medical community, and is expected to help prevent drug interaction errors, over or under treatment, or drug combinations that may cause adverse effects in a patient. Due to these expectations, pharmacists must be well versed in physiology and the actions of drugs and toxins on the human system. A clinical pharmacologist is a pharmacist or other scientist with advanced training and practical experience in the treatment of patients due to an advanced understanding of pharmacokinetics of drugs, as well as clearance issues and metabolic problems that may occur with different drug regimens. Most health care providers depend upon the pharmacologist and pharmacists to assist in patient care to prevent safety issues and provide sound medical care.

A specialized pharmacologist willing to apply their training to the forensic arena is one that has spent considerable time dealing with medication and drug issues and the law, and may be a valuable expert in a case involving alcohol, drugs, or the combination of the two. Many drugs available by prescription, over the counter, and illicitly on the street will have significant adverse effects when mixed with alcohol, and these experts should be able to provide a strong opinion as to how the interaction could have affected the client or victim in a legal matter. It is important to remember that alcohol is a not only a drug, but is the most commonly used and abused drug in the United States. By virtue of the fact that this is a drug, any pharmacologist should be very familiar with alcohol and how it may affect the function of the body, or interact with other drugs or medications. All pharmacists are also familiar with alcohol due to the fact it is used so commonly as a chemical solvent in mixing medications and providing dry chemicals/medications in a liquid preparation.

A licensed and otherwise qualified pharmacologist who is not comfortable in being utilized in a forensic nature should never be used in court unless there is a clear understanding as to what is expected and about what they are expected to provide in their opinion. As with other scientists not particularly well versed in forensics and the application of their expertise to the law, extreme caution should be used in opinions generated by non-forensic pharmacologist.

The forensic physiologist is a scientist that has expertise in the science of physiology and how it may be applied to criminal, civil, or administrative legal issues. Physiology is the study of the functioning of the living systems. It may involve cardiac function, pulmonary function, neurological function, or any other system of the human body. When it involves alcohol physiology, it may actually involve multiple systems due to the whole body aspect of alcohol and its affects. An alcohol physiologist should be knowledgeable in gastrointestinal aspects of the absorption and passage of alcohol to the circulation. There should be considerable knowledge about the impediments to absorption and elimination of alcohol, distribution of alcohol within the body, and the neurological implications of intoxication and impairment. A well trained alcohol physiologist will be able to provide valid opinions about how alcohol is consumed, absorbed, distributed, and eliminated, as well as the effects on the living system to include levels of impairment and how the living body responds to certain levels of alcohol. Any issues are in the arena of the physiologist if that issue stems from the living human system and its interaction with alcohol.

An expert could be qualified as a physiologist by virtue of training and education when that training involves the study of the functioning of the human systems. This could involve a medical doctor, midlevel health care provider (such as a physician assistant or clinical nurse specialist), or other trained health care providers including paramedics with advanced training. Others could qualify by virtue of advanced studies and education to include bachelor's degree studies, master's degree, doctoral degree, or post-doctoral studies in physiology.

Health care providers are particularly qualified in physiology due to the fact that their training *always* involves having to learn the functions of all body systems and pathophysiology and biomedical sciences pertaining to medical training and education. In order to provide health care, and to understand the functioning of the human body at a level that would allow evaluating, diagnosing, and treating different chronic and acute health care issues, one would need a complete understanding of every physiological system and the interactions of these different systems. Couple this knowledge base with the necessary understanding of pharmacology required to safely prescribe medications and you have all of the basics needed to develop a uniquely qualified scientist able to combine all aspects of alcohol and how it may affect the body. In addition, this makes the health care provider one of the few potential forensic experts that requires testing and licensing in every state in the United States. Most states further require intermittent retesting on a regular basis to maintain licensure. This is a clear indication of a scientist that is required to maintain current knowledge in their field, and continually show competence by maintaining licensure under state regulated guidelines.

What is it about these experts that qualify them to provide opinions in legal matters? It is generally required that in order to be qualified as an expert, someone have a combination of experience, education, and training that enables them to have a knowledge about a particular subject beyond that of the normal layperson. There have been challenges in court to expert opinions when a witness is proffered as an expert and they have no doctoral degree. This should easily be overcome by any competent defense attorney, as there is generally no requirement that an expert have a doctoral degree of any type.

There have been challenges to expert opinions when the proffered expert is not a toxicologist. It is a misconception that only a toxicologist can give an opinion as to the effects of alcohol. This probably stems from two issues. Primarily, the state crime labs utilize toxicologist due to the availability, and the fact they are multifunctional in their ability to perform the testing and testify to the results and the effects. In addition, there are more toxicology training programs and certification programs than any other forensic specialty, and this makes them the most visible type of forensic scientist on a regular basis.

An understanding of all the aspects of forensic science clearly provides a view that a scientist should not be disregarded as an expert simply due to the fact that a scientist is not a toxicologist. In fact, there are issues in some cases where the prosecution should not attempt to utilize a toxicologist as they may not be the best expert for the argument at hand. To use the toxicologist on all issues simply because that is what the state has available could jeopardize some cases if the defense vigorously objects to and properly argues against the use of an opinion that may be outside the area of expertise of clinical toxicology.

There have been challenges to some experts due to the fact that they do not have specific training or education in forensic science. Until recently, there have been very few training or degree programs in forensic science, and all forensic scientists have been trained on the job after obtaining their formal education in their particular area of expertise. It is a recent advancement that forensic scientists are formally trained in the forensic aspect of their field. Most forensic experts were scientists first, and gravitated later toward the area of forensic applications of their field of study.

Police departments have many entry level positions in forensic science technology, where a tech can be trained to investigate, collect and package evidence, and present that evidence for scientific evaluation. This technician is certainly a forensic technician even though they may not have a "degree" in forensic science. There is no requirement for an expert in court to have a degree so long as there is adequate training and experience that would allow a potential expert to have more knowledge in a particular matter than the average layperson.

This may change in the future as the field of forensic science is evolving by leaps and bounds and training and degree programs are being developed at a rapid pace around the world. These programs certainly will prepare a scientist for the rigors of developing court room evidentiary presentations and for presenting the material in a valid and professional manner. This should never be used as an argument that the experience and training of the non-formally educated and trained forensic expert may not be up to par. Some of the finest forensic experts in use today do not have degrees in forensic science, but are highly trained and skilled in other fields and through years of experience have become excellent courtroom resources.

When it is determined that an expert may be needed on a case, and the type of expert necessary is selected, it is important to make as much information available to that expert as is possible. Most experts cannot be used effectively if they are only given part of the information required to make a determination of an opinion. An expert opinion is worthless if in the middle of the testimony new information is introduced to the equation and it changes the picture significantly enough to alter an opinion. Surprises of this nature are not welcome to the expert, the attorney, or the court.

Most experts generally will review information on a case for a small fee and will make a determination as to whether they feel there is substantial potential for their opinion to be helpful in your case. No respectable forensic scientist is interested in providing testimony to allow an obviously guilty party to be found innocent of criminal actions. It is important to all ethical forensic experts that their reputation be protected from appearing to be an expert that will whore out their services by providing any opinion because the fee is available. A professional forensic expert will be very selective about the matters in which he is willing to appear in court and provide an opinion. In fact, it is not unusual for an expert to decline many more cases than he would agree to accept.

It is extremely important that an attorney is comfortable with their experts. If an attorney uses a particular expert, and has a comfort level with that expert, he is more likely to understand the reluctance of going into court on a matter in which they have little or no confidence. Although a forensic scientist is not an attorney, and may not understand the legal issues and arguments in a particular issue, they do know the matter from a scientific standpoint. If they are uncomfortable providing an opinion, and can explain that reluctance to the attorney, it is very important that that attorney respect the expert's hesitancy in accepting this case. At no time is it worthwhile to this author as an expert to appear in court and attempt to defend a position in which I really do not believe there may be a valid explanation. At all times, the forensic expert must maintain a professional appearance using an impeccably honest set of facts, and if the case turns on a certain fact, the defense that hired the expert may suffer significantly. No forensic expert worth utilizing will ever risk their reputation to protect an opinion that will damage their standing in court.

It is not unusual for the opposing side to present an expert in rebuttal against the opinion provided by forensic experts. So long as the opinion is based on sound scientific issues and facts, there is little danger of the opinion being altered by opposing expert testimony. It is important to remember that science has different views depending on the area of expertise of the forensic scientist. What is obvious to a physiologist or clinical pharmacologist may not be in the area of expertise of a toxicologist, and there could be what appear to be differences of scientific opinion. This is not to say that one is right and one is wrong, only that there are differences depending on viewpoint and areas of expertise. It is very important to understand these differences in order to provide logical legal arguments to the opinions and how they should be applied. This is what makes expert scientist types so important when choosing the expert to hire.

In a recent case in a court in central Virginia, there were actually two defense experts, one a pharmacologist with extensive research and writing experience in alcohol and its issues as to absorption and elimination rates, as well as testing and the potential differences in blood and breath results. His experiences were exceptional in that he held an associate professorship in a prestigious medical school, and instructed medical, pharmacy, and allied health students in exactly this information on a daily basis. He had extensive experience testifying in court on similar matters. The second defense expert was a physiologist, with credentials to practice medicine (as a physician assistant), and experience as a police officer skilled in DUI enforcement prior to his medical training. He held a degree in biology from a large respected university, as well as his training from a highly regarded medical school for his PA education and clinical training. He also had extensive experience in testifying in court for many years on this same subject matter.

The prosecution was prepared with an expert from the Department of Forensic Sciences in Richmond. Her background was that as a toxicologist, although her degree was not in toxicology. She did hold a bachelor degree in analytical chemistry, and a masters degree in forensic science. Her current employment was with the Commonwealth of Virginia, as the supervisor of the breath testing section of the state lab. Her position is to train the law enforcement community in the use of the current breath testing instrumentation. She described her knowledge in DUI issues as obtained from reading scientific journals on the subject, and attending a national seminar on DUI testing. She took a single 3 semester hour course in pharmacokinetics in her educational experience. She stated under defense questioning that she had no clinical experience in medicine or diagnosing health conditions, nor did she have any background in clinical medicine issues.

First, it is important to state up front, before dissecting this matter, the state's expert is exceptionally well versed and highly qualified on the Intoxilyzer 5000 and breath testing. She is bright, very knowledgeable, and presents well in court. She is honest, and the courts rightfully have the highest respect and regards for this scientist. When comparing these three scientists, it is clear that there are significant differences in their expertise, and their levels of education and experience. There was no question that they all three had different areas of expertise and training.

The issue on this particular matter was gastroesophageal reflux disease, and whether the certificate's results could have been overstated for the breath alcohol levels over the actual blood alcohol levels. The two defense experts testified extensively about GERD, and how it could allow an error in the Intoxilyzer 5000 to give a falsely elevated alcohol reading. Both testified that the drinking history supplied by the defendant was not consistent with the reading on the certificate, and with the documented history of GERD in the defendant, it was possible that the reading was in error. The Commonwealth's expert testified in the same manner and stated the drinking history was not consistent with the reading presented on the certificate. There was no disagreement between all three experts on this issue.

The variance came in when the Commonwealth's expert gave testimony that her opinion was that ***the defendant was being untruthful as to the alcohol consumption***. She refused to accept the possibility of any other potential source for this discrepancy. She further testified that it was her understanding that the slope detector in the instrument currently in use would eliminate any mouth alcohol that may be present, even though both defense experts testified extensively that mouth alcohol and alcohol vapors present from severe GERD were not the same and the slope detector absolutely could miss the alcohol present from GERD due to the lack of down slope after the initial spike as seen in mouth alcohol. Both testified extensively as to personal observations during research of failure on behalf of the slope detector to notify the Intoxilyzer operator of refluxed alcohol. These were repeatable and predictable results.

The Commonwealth's expert stated during her testimony that they were taught not to calculate possible BAC's as there were many variables involved, but that they should always take a known value, that on the certificate, and calculate back from that value, using an elimination rate of 0.015% to 0.020%, and considering the fact that they always assume the absorption is complete in 30 minutes, they arrive at a higher BAC at the time of the driving by several points. They consider no other facts than those in their "expert" opinion. With all due respect to a scientist that I believe to be highly qualified and respected, I believe there are several glaring errors in the factorial presentation of the state's experts in these matters.

As a scientist, I find it difficult to accept that a colleague could go into court and present an opinion as biased as one that would assume only that the defendant is lying and that there is no other possibility than the scenario that the defendant had his highest BAC at (or before) the time of the driving and came down to the level indicated by the certificate of analysis. This opinion requires the certificate to be considered fully accurate in order to even present this argument. This is very narrow minded and refuses to acknowledge volumes of research that have been documented over the last half century that indicates multiple variables in absorption, elimination, and clearance rates of alcohol. As a scientist testifying in court, I would never consider testifying to the fact that one scenario and only one scenario was possible. It is important to make it clear that there were many possibilities, and the point of scientific testimony is to make those potentialities clear to the trier of fact in a case.

Certainly there were other factors involved in this particular case as there are in all cases. Driving behavior must be considered and performances on the standardized field sobriety tests are issues to be evaluated. I found it interesting in this particular case that driving was not terribly abnormal, but was essentially excess speed, movement off the pavement into a grass edge by "1/2 the width of the tire", and driving within three feet of a uniformed police officer in a navy blue uniform, in the dark, on a narrow road way with a large number of cars parked along one side (partially on the road!), and two pedestrians present near the parked cars. Not perfect driving, and possibly driving indicative of only someone trying to give a wide berth to parked vehicles, but certainly not indicative of someone with a BAC in the range of one who would be obviously drunk. Field sobriety testing consisted of a battery of tests that are in no way a part of the NHTSA battery of standardized tests, but were accepted as "informative" by the court. Most of these tests were evaluated by the NHTSA and found to be useless for determination of impairment.

The importance of this particular case is the difference of the three experts. Even though all had different backgrounds, and different areas of expertise, it appeared the Commonwealth's expert was given considerable weight, even though she clearly didn't have the area of expertise of the other two experts in the primary issue at hand (GERD). She had more latitude in being permitted to testify about a subject with which she admittedly had no knowledge. Her area of expertise was instrumentation, and that is the area in which she is employed by the state. Had her testimony been limited to that area, then there would have been no opposing testimony to the two defense experts. Since the prosecuting attorney needed a rebuttal expert, he proffered this expert since she was there as an agent of the Commonwealth and successfully entered her testimony even with her own admitted minimal experience and education on the topic of reflux.

In light of the driving behavior and lack of failure on the Standardized Field Sobriety Tests, the evidence was quite supportive of the opinions of both defense experts. Remember, neither defense expert testified that the alcohol reading on the Intoxilyzer came from GERD, only that there was a significant chance. Should this have led to a "reasonable doubt" as argued by the defense counsel?

At no time should any forensic expert be forced to testify in any matter. This author has had the unfortunate experience of being forced to testify in a criminal matter by subpoena. The result was of course disastrous for the defendant that hired me, as well as some damage to my own reputation in this particular court. I attempted to strongly discourage the attorney wishing to hire my services from forcing me to testify, but he was adamant that I appear. I felt very strongly against his client (not personally, but in the scientific merit of his case) and the opinion in which they were attempting to elicit in this matter, and it was clear in court. If your expert advises against using their services on a case, respect this fact, and either find a different expert, or advise your client that they do not have the strength of argument for a valid scientific defense. This is where it is so very important to have a pool of forensic experts in which you have confidence and are willing to trust when they offer an opinion or refuse to offer an opinion.

I was involved in one case where the client clearly was changing the facts of a case after I had evaluated the information. My opinion was favorable to the client, and they retained my services to appear in court. The matter involved considerable travel and time expenditure, but after appearing at the Courthouse on the morning of trial, the client completely changed the facts upon which I had formulated my opinion. At that time, the opinion was no longer satisfactory to the client, and the client offered to alter the facts back to the original set of numbers. I immediately withdrew from this matter, with the approval of the attorney. The client was extremely upset with both the attorney and with me, and made quite a scene in the hallway of the courthouse. At that point, there was no possibility of me appearing on behalf of this client, and the attorney continued the matter for the convenience of the client in order to determine if there was any other valid mechanism with which to attempt to provide a defense. This was a very ugly matter, and lasted for several months afterward, but fortunately, the attorney was very willing to protect me from the client due to a full understanding of the role of an expert.

Any attorney that considers hiring an expert in any matter needs to ensure the client has a full, complete understanding of the role of a defense expert. It is imperative that the client considering an expert must understand this role and the expert's intended involvement in the case. At no time should any expert fee be based on outcome at trial. The expert is never paid for what he is to say, or for his opinion, as the facts presented at trial will always guide the expert in his calculations, considerations, and opinions. The expert is paid for his time, and the time it takes to prepare for trial. This is what every client needs to be advised that he or she is purchasing when retaining an expert for court appearances. Frequently there may be a small fee for the initial consultation, but the majority of the fee will involve the time and expense of travel to appear at trial, as well as time spent in court waiting to testify. It is rare that a specific time can be given to an expert to appear, and this time must be considered when fees are set.

All expert witnesses should make the fee and/or fee structure clear at the outset. There should be no surprises to the attorney or client in the amount of fee set for preparation and appearance at trial. I personally have developed a set fee over the years. I previously had a retainer, and charged an hourly rate against that retainer. There were times when the client ended up with money in return, but more often than not, there was more owed in the end. To prevent this, I found a set fee for a single day of appearance, travel, and two hours of preparation time to be adequate and fair. When this was evaluated over time, I found it was normally very close to the actual hourly rate, and it eliminated any concerns over collecting additional funds after the trial, or the client wondering about any return fee after the fact. I have found this to be a fair and equitable method of charging for my services and for the most part have stuck with this set fee plan. Occasionally, if there are to be more than one day of trial, and (or) extensive preparations, there may be additional fees, but this can normally be determined ahead of time and made clear to the client and attorney.

No expert should consider discounting fees for volume from an attorney, as this may possibly indicate some impropriety between the attorney and the expert. All possible questions of favoritism should be eliminated in order to ensure there is no question involved in a case other than the issue under consideration. I will occasionally discount a fee if there is a genuine difficulty in a client's ability to pay. My fees are not inexpensive, and should a client have a definite scientific defense, I would never allow lack of funding to interfere with their ability to present all the evidence necessary to obtain a fair trial.

When looking at potential experts in a matter, the expert should be selected because he is able to evaluate the facts and present as concise an opinion as possible. If there is additional information necessary, most reputable experts would not hesitate to with hold the opinion until all the necessary information is available. This is good, sound practice and should never be considered as being difficult or misleading. Should information become available in a matter after the opinion has been disclosed, then that information should always be provided to the expert. As mentioned earlier, the witness stand is not the place for the expert to learn additional facts in any case.

Should the facts of the matter change, the attorney should be very clear with the client and the expert about the effects of these changes. Clients need to be kept informed as to potential damage certain information may have in their case. Be very wary of any expert who may be willing to avoid certain facts, as that expert clearly is not interested in the dispensing of appropriate justice, and may damage your case should those facts trap the expert on the stand.

Most experts are willing to discuss the matter with opposing attorneys, but this should always be a decision of the attorney that has retained the expert. If the facts are solid, and there is little argument as to the facts, it can sometimes encourage an acceptable plea arrangement in the matter, and a trial may be avoided completely. Some clients are not interested in potential plea arrangements, but they should always be willing to discuss potential limitation of exposure to the system. No expert can ever give any type of guarantee to a client as to the acceptance of their testimony by the court. Each case is different, and past history is seldom a definite indication of outcome in any matter. The same judge, prosecutor, police officer, and similar facts may turn out one way on one occasion, and completely different on another.

If a reasonable plea arrangement can be obtained, that is the only real guarantee in any case. To take any matter to actual trial constitutes a gamble, and the presentation of facts and opinions may fall into disarray depending upon the performance of the opposing attorneys. What appears to be very clear and concise scientific argument before trial may or may not come out fully depending upon the acceptance of the testimony by the court. This is not to say that a plea agreement should always be sought and/or taken in every matter, but any pre-trial agreement that is reasonable and dispenses the appropriate level of justice should certainly be considered; otherwise, there is no reason to not present the evidence and trust the judgment of the court with all of the factors in the open.

Afterword

This text was born of the idea that DUI enforcement has become such a scientific offense that there is considerable room for error due to lack of full understanding in the science behind this issue. Nearly every law enforcement officer I have ever met is interested only in protecting the public from dangerous individuals. There has never been anyone wearing a badge and tasked with protecting the public that I have known that would ever be willing to put an innocent person in jail for any reason.

This text is not meant to anger, demean, belittle, or lessen the importance of any law enforcement officer in any way. The same goes for prosecuting attorneys, judges, and state forensic experts. I have the greatest respect for each and every player in this complex process.

Hopefully, everyone that reads this manual will understand that this is nothing more than an attempt to assist each and every person involved in catching, trying, and incarcerating drunk drivers. I have been dismayed over the years by the occasional overzealous prosecutor that absolutely refuses to accept the fact that a person arrested for a crime may not be guilty, for what ever reason. It is natural that every law enforcement officer that arrests anyone believes completely in the fact that he or she has arrested someone guilty of committing a crime. No officer would arrest anyone when he or she honestly thought the person was innocent. It is a good situation when the prosecutors are willing to pull all the stops to prove the officer correct.

There may well come a time when it becomes clear that there could be reasonable doubt as to the guilt of an arrested citizen, and it's at that time that everyone needs to step back and re-look at the facts. To come to the conclusion that someone may be able to inject this reasonable doubt demands that the case be properly taken to the correct ending. This in no way should reflect on any officer or prosecutor, as probable cause to make an appropriate arrest is far different than beyond a reasonable doubt necessary for a criminal conviction.

Nearly each and every Commonwealth's expert I have encountered over the years has honestly believed in their opinion. They have been professional, very well trained, and willing to spend what ever time is necessary to help the court understand the issues from their viewpoint.

I have participated in trials where I have become the person on trial. I have felt that there were times when the prosecution team could not counter my message on a particular case, and therefore did the next best thing for their team and attacked me, the messenger. On occasion I have been forced to defend myself before being able to provide what was obvious from a scientific standpoint. It became sad that the actual issue was secondary to the trial attacking the experts. What was even more troubling for me was the fact that the courts were usually quite adamant about not allowing the defense team to be aggressive in questioning state's experts, but would sit back and allow prosecutors to literally abuse me on the witness stand. On the other hand, there were certainly many judges that protected me just as much as the state's experts, even when the defense attorney that brought me to court sat back and did nothing to stop an overly aggressive prosecutor.

Hopefully, this will enlighten enough members in the judicial process to prevent this type of problem in the future. I believe that every prosecutor and judge has an incredibly difficult job. There is frequently a very fine line between guilt and innocence, and I would never want to be in the position to make that decision. I am thankful for those of you who are willing to make a lifelong career out of providing this necessary service. I have seen many judges literally torn over making this decision, but they come back day after day to sit in judgment of others. This is an amazing weight on the shoulders of men and women that are all too human themselves.

Every person in the United States that is accused of a crime and arrested has the right to present the best defense. It is the duty of our legal system to ensure that anyone can properly defend their actions in court. I don't believe it is appropriate to take the obvious guilty party and grab at technicalities to limit having to take proper responsibility for their actions. Most defense experts that I have met feel the same way, and very few would ever appear in court if they believed that the person attempting to retain their services was actually guilty.

It is not my goal to turn drunk drivers loose on the street. This is not in the best interest of anyone, including my friends and family, and dangerous drunks need to be removed from the driving public. That is, and will continue to be, the goal of everyone involved in arresting, trying, and punishing these dangerous members of society.

I wish good luck to everyone involved in this endeavor, and I hope you catch them all.

Alfred W. O'Daire, Jr.

End Notes

[1] There are other limitations that need to be addressed with this "slope detector" that are addressed elsewhere.

[2] "Reliability of Breath-Alcohol Analysis in Individuals with Gastroesophageal Reflux Disease", J Forensic Sci, 1999; 44(4): 814-818; Alan Wayne Jones, Ph.D, D.Sc., and others

[3] ibid

[4] *Gastric reflux, Regurgitation, and the Potential Impact of Mouth Alcohol on the Results of Breath Alcohol Testing*; Jones, Alan Wayne, PhD, DSc, 16 August 2006

[5] *Final report—Validation of SFST Battery at BACs Below 0.10 Percent* (Stuster & Burns, 1998)

[6] Id